More Praise for Mary Ro

"Science writing doesn't get funnier or more human than this." —Marta Salij, *Detroit Free Press*

"Ms. Roach's gift for facetiousness serves her well here. *Spook* is dependably witty, especially when it ventures far into the ether. . . . And it is populated by vividly evoked oddballs. . . . Ms. Roach makes herself a wry, enjoyable character throughout the book's escapades . . . a clever investigator and a thoroughly entertaining, if skeptical, tour guide."
—Janet Maslin, *New York Times*

"Funny and smart . . . since she's a scientist at heart, she also lasers through the smoke and mirrors." —*People*

"First, there's her wit and stylistic brio. From the clever dedication . . . to her gift for terse summation . . . to her genuine humility . . . Roach is a clear and versatile writer. She has a sharp eye for detail that demonstrates her traditional journalistic skills . . . but she delivers her findings in ultracontemporary tones. . . . She has a huge heart, a strong sense of empathy for the oddball, and she's willing to go to great lengths to find and report stories from the hinterlands of understanding."
—Floyd Skloot, *Chicago Tribune*

"Roach is a wonderfully vivid writer . . . [with] a keen eye for the perfect detail, an ear for the zinging quotation and a finely tuned sense of the preposterous. . . . A celebration of the wide, occasionally crazy spectrum of human pursuit."
—Kate Zernike, *New York Times Book Review*

"Sly . . . irreverent . . . and downright witty. . . . Roach wields the scientific method and her skeptical mind like surgical tools, cutting away at myth, pseudo-theory and madness until all that remains is fact. In the end, believers and non-believers will be satisfied by Roach's conclusions. . . . Reading *Spook* is like attending a lecture by a professor who is equal parts Groucho Marx and Stephen Jay Gould, both enlightening and entertaining. Roach makes learning about anything, be it decomposing bodies or the possibility of a verifiable afterlife, pleasurable." —Dorman T. Schindler, *Sunday Denver Post & Rocky Mountain News*

"This is Roach at her best." —*San Francisco Magazine*

"A sharp-eyed supernatural history."
—Cathleen Medwick, *O Magazine*

"Funny, inquisitive and uncowed by experts, she's the general reader's ideal emissary to the arcana of serious science. . . . Roach's writing has what science has so far failed to find: a divine spark." —Malcom Jones, *Newsweek*

"Investigative reporting has no lighter, more irreverent spirit than Mary Roach. . . . What lets *Spook* rise above the dry survey of (mostly inconclusive) scientific investigations it could have been is Mary Roach—her lively and distinctive style, or perhaps more accurately put, her attitude. . . . Roach is funny, fair-minded, impartial and endlessly curious. . . . *Spook* is enormous fun." —David A. Walton, *Pittsburgh Union-Tribune*

"*Spook* is nothing if not amusing. Roach heads into a mire of ghostly and laboratory episodes with robust humor."
—Nora Seton, *Houston Chronicle*

"Oh, had Roach only been my high school science teacher. *Spook* is filled with fascinating characters, wacky experiments, and Roach's accessible scientific reporting paired with comic relief and gentle insight."

—Brooke Gilbert, Amazon.com

"Roach is a self-described skeptic, but one with an open mind, a sense of adventure and a ready quip. All of this makes her an amiable and entertaining guide as she traverses several continents to look for scientific proof for the great beyond."

—Megan Harlan, *San Francisco Chronicle*

"Alas, she doesn't find the answers. But Roach is such a smart and breezy companion that it's enough to watch her realize that in the end she might not need them."

—Karen Valby, *Entertainment Weekly*

"As fascinating and thorough as her research may be, the greater tale lies in the people: The mediums, the mystics, the I-want-to-believers, the scientists, pseudo and for-real. Roach has a genius for portraiture, and she can bring the oddest people to hilarious life without a hint of condescension . . . her oddball, incandescent personality radiates from her prose."

—Arthur Salm, *San Diego Union Tribune*

"*Spook* is filled with some mind-blowing ideas that will make you glad you've got an open mind."

—Chris Watson, *Santa Cruz Sentinel*

"Roach brings to *Spook* a lightness and a sense of humor that, happily, smooth the morbid edges of the proceedings she describes. . . . The most refreshing thing about *Spook* is that Roach herself is a skeptic, guiding a skeptic's tour. . . . What evidence she does come across, therefore, becomes all the more compelling."

—Priya Jain, *Salon*

"Short of the Ultimate Trip (the one with the light and the pearly gates), it's about as entertaining a journey out of the realm of the living as anybody could want."

—Donna Bowman, *The Onion*

"Her biting wit is omnipresent from the start. . . . This is one fantastically enjoyable book." —curledup.com

"For all Roach's skeptical and often hilarious accounts, she is an eager volunteer and ready to accept evidence if evidence there be. . . . Throughout, she is critical and witty [and] truly deft handling of the (mostly) daft." —*Kirkus Reviews*

"She has done it again. . . . [Roach] now presents an equally smart, quirky, hilarious look at whether there is a soul that survives our physical demise. Roach perfectly balances her skepticism and her boundless curiosity with a sincere desire to know. . . . An original who can enliven any subject with wit, keen reporting and a sly intelligence." —*Publishers Weekly*

"Roach is dogged in her approach. . . . Gripping . . . Roach's witty asides liven up an already interesting and unusual read."

—*Booklist*

"Science writer Mary Roach's wit and flair for vivid storytelling . . . have earned her a loyal readership, and her new book will only cement it." —*Ruminator Review*

"It's a fabulous read, both a comic and a serious investigation of the history of séances, mediums, spirits and discarnate voices. . . . All of a sudden, with *Spook*, the field of the paranormal is bust-a-gut funny. . . . *Spook* is a comic romp through a mix of history and the current practices of a particular culture." —Monica Drake, *Sunday Oregonian*

"Roach's humorous scoffings will make even the most adamantly-believing readers chuckle. . . . No matter what you believe, pick up a copy of *Spook*."
—*Vail Trail*

"*Spook* is a hilarious look at misadventures in paranormal research. . . . In the sharp-witted world of Mary Roach, the answer is inconsequential. The interesting part is the question itself—and the eccentric characters doing the asking. . . . Surreal, fascinating, at times absurd and always hilarious, Mary Roach may not reveal the street address of our final destination, but in *Spook* she makes it sound less like a morgue and more like a comedy club."
—Vince Darcangelo, *Boulder Weekly*

SPOOK

ALSO BY MARY ROACH

Gulp: Adventures on the Alimentary Canal

Packing for Mars: The Curious Science of Life in the Void

Bonk: The Curious Coupling of Science and Sex

Stiff: The Curious Lives of Human Cadavers

W. W. NORTON & COMPANY
NEW YORK | LONDON

SPOOK

Science Tackles the Afterlife

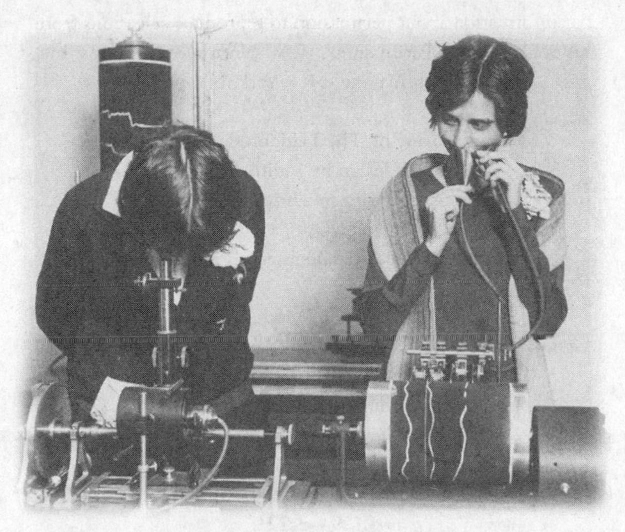

Mary Roach

Printed in the United States of America
First published as a Norton paperback 2006

Photograph credits: title page: Underwood & Underwood/Corbis; p. 23: Getty Images/Robert Holmgren; p. 55: Getty Images/Derek Berwin; p. 77: Hulton-Deutsch Collection/Corbis; p. 107: Getty Images/Digital Vision; p. 121: Mary Evans Pictures Collections; p. 149: Getty Images/Wallace Kirkland; p. 169: Getty Images/Andrea Pistolesi; p. 179: H. Armstrong Roberts/Corbis; p. 213: Hulton-Deutsch Collection/Corbis; p. 225: Getty Images/Andrea Chu; p. 239: courtesy of Grant Sperry; p. 261: Bettman/Corbis

Manufacturing by The Haddon Craftsmen, Inc.
Book design by Judith Abbate
Production manager: Amanda Morrison

Library of Congress Cataloging-in-Publication Data

Roach, Mary.
Spook : science tackles the afterlife / by Mary Roach.—1st ed.
p. cm.
Includes bibliographical references.
ISBN 0-393-05962-6 (hardcover)
1. Future life. 2. Religion and science. I. Title.
BL535.R63 2005
129—dc22
2005014450

ISBN-13: 978-0-393-32912-4 pbk.
ISBN-10: 0-393-32912-7 pbk.

W. W. Norton & Company, Inc., 500 Fifth Avenue, New York, N.Y. 10110
www.wwnorton.com

W. W. Norton & Company Ltd., Castle House, 75/76 Wells Street, London W1T 3QT

0

FOR MY PARENTS,
WHEREVER THEY ARE OR AREN'T

Contents

Introduction 11

1. You Again
A visit to the reincarnation nation 23

2. The Little Man Inside the Sperm, or Possibly the Big Toe
Hunting the soul with microscopes and scalpels 57

3. How to Weigh a Soul
What happens when a man (or a mouse, or a leech) dies on a scale 79

4. The Vienna Sausage Affair
And other dubious highlights of the ongoing effort to see the soul 109

5. Hard to Swallow
The giddy, revolting heyday of ectoplasm 123

6. The Large Claims of the Medium
Reaching out to the dead in a University of Arizona lab 151

7. Soul in a Dunce Cap
The author enrolls in medium school 171

8. Can You Hear Me Now?
Telecommunicating with the dead 181

9. Inside the Haunt Box
Can electromagnetic fields make you hallucinate? 215

10. Listening to Casper
A psychoacoustics expert sets up camp in England's haunted spots 227

11. Chaffin v. the Dead Guy in the Overcoat
In which the law finds for a ghost, and the author calls in an expert witness 241

12. Six Feet Over
A computer stands by on an operating room ceiling, awaiting near-death experiencers 263

Last Words 293

Acknowledgments 297
Bibliography 299

Introduction

MY MOTHER worked hard to instill faith in me. She sent me to catechism classes. She bought me nun paper dolls, as though the meager fun of swapping a Carmelite wimple for a Benedictine chest bib might inspire a taste for devotion. Most memorably, she read me the Bible. Every night at bedtime, she'd plow through a chapter or two, handing over the book at appropriate moments to show me the color reproductions of parables and miracles. The crumbling walls of Jericho. Jesus walking atop stormy seas with palms upturned. The raising of Lazarus—depicted in my mother's Bible as a sort of Boris Karloff knockoff, wrapped in mummy's rags and rising stiffly from the waist. I could not believe these things had happened, because another god, the god who wore lab glasses and knew how to use a slide rule, wanted to know how, scientifically speaking, these things could be possible. Faith did not take, because science kept putting it on the spot. Did the horns make the walls fall, or did there happen to be an earthquake while the priests were trumpeting? Was it possible Jesus was making use of an offshore atoll, the tops of which sometimes lie just inches below the water's surface? Was Lazarus a simple case of premature entombment? I wasn't saying these things didn't happen. I was just saying I'd feel better with some proof.

Of course, science doesn't dependably deliver truths. It is as fallible as the men and women who undertake it. Science has the answer to every question that can be asked. However, science reserves the right to change that answer should additional data become available. Science first betrayed me in the early eighties, when I learned that brontosaurus had lived in a sere, rocky desert setting. The junior science books of my childhood had shown brontosaurs hip-deep in brackish waters, swamp greens dangling from the sides of their mouths. They'd shown tyrannosaurs standing erect as socialites and lumbering Godzilla-slow, when in reality, we were later told, they had sprinted like roadrunners, back flat and tail aloft. Science has had us buying into the therapeutic benefits of bloodletting, of treating melancholy with arsenic and epilepsy with goose droppings. It's not all that much different today: Hormone replacement therapy went from miracle to scourge literally overnight. Fats wore the Demon Nutrient mantle for fifteen years, then without warning passed it to carbohydrates. I used to write a short column called "Second Opinion," for which I scanned the medical literature, looking for studies that documented, say, the health benefits of charred meat or the deleterious effects of aloe on wound-healing. It was never hard to fill it.

Flawed as it is, science remains the most solid god I've

got. And so I decided to turn to it, to see what it had to say on the topic of life after death. Because I know what religion says, and it perplexes me. It doesn't deliver a single, coherent, scientifically sensible or provable scenario. Religion says that your soul goes to heaven or possibly to a seven-tiered garden, or that your soul is reincarnated into a new body, or that you lie around in your coffin clothes until the Second Coming. And, of course, only one of these can be true. Which means that for millions of people, religion will turn out to have been a bum steer as regards the hereafter. Science seemed the better bet.

For the most part, science has this to say: Yeah, *right.* If there were a soul, an etheric disembodied you that can live on, independent of your brain, we scientists would know about it. In the words of the late Francis Crick, codiscoverer of the structure of DNA and author of *The Astonishing Hypothesis: The Scientific Search for the Soul*, "You, your joys and your sorrows, your memories and your ambitions, your sense of personal identity and free will, are in fact no more than the behavior of a vast assembly of nerve cells and their associated molecules."

But can you *prove* that, Dr. Crick? If not, then it's no more good to me than the proclamations of God in the Old Testament. It's just the opinion, however learned, of one more white-haired, all-knowing geezer. What I'm after is proof. Or evidence, anyway—evidence that some form of disembodied consciousness persists when the body closes up shop. Or doesn't persist.

Proof is a tremendously comforting thing. When I was little, I used to worry that one day, without warning, the invisible forces that held me to the earth were going to conk out, and that I would drift up into space like a party balloon, rising and rising until I froze or exploded or suffocated or all three at once. Then I learned about gravity, the dependable pull of the very large upon the very small. I learned that it had been scientifically proven to exist, and I no longer worried about floating away. I worried instead about blackheads and whether Pat Stone dreamed of me and other dilemmas for which science could provide no succor.

It would be especially comforting to believe that I have the answer to the question, What happens when we die? Does the light just go out and that's that—the million-year nap? Or will some part of my personality, my me-ness, persist? What will that feel like? What will I do all day? Is there a place to plug in my laptop?

Most of the projects that I will be covering have been—or are being—undertaken by science. By that I mean people doing research using scientific methods, preferably at respected universities or institutions. Technology gets a shot, as does the law. I'm not interested in philosophical debates on the soul (probably because I can't understand them). Nor am I going to be relating anecdotal accounts of personal spiritual experiences. Anecdotes are interesting, occasionally riveting, but never are they proof. On the other hand, this is not a debunking book. Skeptics and debunkers provide a needed service in this area, but their work more or less assumes an outcome. I'm trying hard not to make assumptions, not to have an agenda.

Simply put, this is a book for people who would like very much to believe in a soul and in an afterlife for it to hang around in, but who have trouble accepting these things on faith. It's a giggly, random, utterly earthbound assault on our most ponderous unanswered question. It's spirituality treated like crop science. If you found this book in the New Age section of your local bookstore, it was grossly misshelved, and you should put it down at once. If you found it while browsing Gardening, or Boats & Ships, it was also misshelved, but you might enjoy it anyway.

AUGUST 6, 1978, was a Sunday, the Feast of the Transfiguration. It was evening, and Pope Paul VI lay dying in his bedroom. With him was his doctor and two of his secretaries, Monsignor Pasquale Macchi and Father John Magee. At 9:40 p.m., following a massive heart attack, His Holiness expired. At that very moment, the alarm clock on his bedside table rang out. Accounts of this episode refer to the timepiece as the Pope's "beloved Polish alarm clock." He bought it in Warsaw in 1924 and carried it with him in his travels from then on. He seemed to be fond of it in the way that farmers are fond of old, slow-moving dogs, or children of their blankets. Every day, including the day he died, the alarm was set for 6:30 a.m.

I first came upon this story in a gullible and breathless compilation of supposed evidence for the afterlife. I don't recall the book's title (though the title of the chapter about spirit communication—"Intercourse with the Dead"—seems to have stayed with me). The book presented the story of the pontiff's noisy passing as proof that some vestige of His Holiness's spirit influenced the papal clockworks* as it departed the body. *Pontiff,*

*A modern corollary to the Pope's alarm clock can be found in the erratic behavior of a digital alarm clock belonging to a Mrs. Linda G. Russek, of Boca Raton, Florida. Russek's husband Henry had recently died, and she wondered whether he was trying to communicate with her via the clock. Russek, a parapsychologist, undertook an experiment in which she asked Henry to speed the clock up on even days and slow it down on odd days. Alas, the data were meaningless because shortly after the study began, the AM/PM indicator had gone on the blink, and Russek was unable to definitively conclude anything beyond the fact that she needed a new alarm clock.

a popular biography of Paul VI, relates the tale with similar cheesy dramatics: "At that precise moment the ancient alarm clock, which had rung at six thirty that morning and which had not been rewound or reset, begins to shrill. . . ."

In Peter Hebblethwaite's *Paul VI: The First Modern Pope* we find a different take on the proceedings. In the morning of his last day, the Pope is sleeping. He awakes and asks the time and is told it's 11 a.m. "Paul opens his eyes and looks at his Polish alarm clock: it shows 10:45. 'Look,' he says, 'my little old clock is as tired as me.' Macchi tries to wind it up but confuses the alarm with the winder." By this version, the alarm went off at the moment the Pope died because Monsignor Macchi *had accidentally set it for that moment.*

I am inclined to side with Hebblethwaite, because (a) his book is studiously footnoted and (b) Hebblethwaite doesn't gild his renderings of papal life. For instance, we have the scene in the final chapter wherein Pope Paul VI is lying in bed watching TV. Not only is the earth's highest-ranking Catholic, the Holy of Holies, watching a B-grade western, he is having trouble following it. Hebblethwaite quotes Father Magee, who was there at the time: "Paul VI did not understand anything about the plot, and he asked me every so often, 'Who is the good guy? Who is the bad guy?' He became enthusiastic only when there were scenes of horses." Hebblethwaite tells it like it is.

Just to be certain, I decided to track down the man who either did or didn't mess with the winder: Pasquale Macchi. I called up the U.S. Conference of Catholic Bishops, the American mouthpiece of the Catholic Church, and was put through to the organization's then librarian, Anne LeVeque. Anne is an accommodating wellspring of Catholic-related trivia, including the stupendously odd fact that freshly dead popes are struck thrice on the forehead with a special silver hammer. LeVeque knew someone in the organization who had spoken with a group of priests who had met with Macchi shortly after Paul VI's death, and she gave me his number. He agreed to tell me the story, but he would not reveal his name. "I'm better as your Deep Throat," he said, forever linking in my head the U.S. Conference of Catholic Bishops with porn movies, a link they really and truly don't need.

Deep Throat confirmed the basic story. "It was described to me as not instantaneous, but more of a five, four, three, two, one . . . and the alarm went off." He was told that despite what others had said, the clock had not been set for the time at which Paul died. "The feeling," he said, "was that what it suggested was the departure of Paul VI's soul from his body." Then he looked up Macchi's address for me, in what he called the Pontifical Phone Book. I wanted to ask if it included a Pontifical Yellow Pages, with pontifical upholstery cleaners and pontifical escort services, but managed not to.

Macchi is a retired archbishop now. With the help of a friend's friend from Italy, I drafted a note asking about the alarm clock incident. Archbishop Macchi wrote back promptly and courteously, addressing me as "Gentle Scholar," despite my having addressed him as Your Eminence (suggesting mere cardinalhood) when in fact he is either a Your Excellency or a Your Grace, depending on whose etiquette book you consult. (Your Holiness, reserved for the Pope himself, trumps all, except possibly, in my hometown anyway, Your San Francisco Giants.) Macchi included a copy of his own biography of Paul VI, with a bookmark at page 363. "In the morning of that day," he wrote, "having noticed that the clock was stopped, I wanted to wind it up and inadvertently I had moved the alarm hand setting to 9:40 p.m." Deep Throat's deep throats, it seems, had led him astray.

Annoyingly, I came across yet a third version of the alarm clock incident, this one by a priest with a grudge against Paul

VI. This man held that the clock story had been fabricated by the Vatican as evidence for a false time of death, part of an effort to cover up some breach of papal duty that would have made the Pope seem impious.

The moral of the story is that proof is an elusive quarry, and all the more so when you are trying to prove an intangible. Even had I managed to establish that the alarm clock had indeed gone off for no obvious mechanical reason at the moment the pontiff died, it wouldn't have proved that his departing soul had triggered it. But I couldn't even get the clock to stand and deliver.

The deeper you investigate a topic like this, the harder it becomes to stand on unshifting ground. In my experience, the most staunchly held views are based on ignorance or accepted dogma, not carefully considered accumulations of facts. The more you expose the intricacies and realities of the situation, the less clear-cut things become.

And also, I hold, the more interesting. Will I find the evidence I'm looking for? We'll just see. But I promise you a diverting journey, wherever it is we end up.

SPOOK

You Again

A visit to the reincarnation nation

I DON'T RECALL my mood the morning I was born, but I imagine I felt a bit out of sorts. Nothing I looked at was familiar. People were staring at me and making odd sounds and wearing incomprehensible items. Everything seemed too loud, and nothing made the slightest amount of sense.

This is more or less how I feel right now. My life as a comfortable, middle-class American ended two nights ago at Indira Gandhi International Airport. Today I am reborn: the clueless, flailing thing who cannot navigate a meal or figure out the bathrooms.

I am in India spending a week in the field with Kirti S. Rawat, director of the International Centre for Survival (as in survival of the soul) and Reincarnation Researches. Dr. Rawat is a retired philosophy professor from the University of Rajasthan, and one of a handful of academics who think of reincarnation as something beyond the realm of metaphor and religious precept. These six or seven researchers take seriously the claims of small children who talk about people and events from a previous life. They travel to the child's home—both in this life and, when possible, the alleged past life—interview family members and witnesses, catalogue the evidence and the discrepancies, and generally try to get a grip on the phenomenon. For their trouble, they are at best ignored by the scientific community and, at worst, pilloried.

I would have been inclined more toward the latter, had my introduction to the field not been in the form of a journal article by an American M.D. named Ian Stevenson. Stevenson has investigated some eight hundred cases over the past thirty years, during which time he served as a tenured professor at the University of Virginia and a contributor to peer-reviewed publications such as *JAMA* and *Psychological Reports*. The University of Virginia Press has published four volumes of Stevenson's reincarnation case studies and the academic publisher Praeger recently put out Stevenson's two-thousand-page opus *Biology and Reincarnation*. I was seduced both by the man's credentials and by the magnitude of his output. If Ian Stevenson thinks the transmigration of the soul is worth investigating, I thought, then perhaps there's something afoot.

Stevenson is in his eighties and rarely does fieldwork now. When I contacted him, he referred me to a colleague in Bangalore, India, but warned me that she would not agree to anything without meeting me in person first—presumably in Bangalore, which is a hell of a long way to go for a get-acquainted chat. A series of unreturned e-mails seemed to confirm this fact. At around the same time, I had e-mailed Kirti Rawat, whom Stevenson worked with on many of his

Indian cases in the 1970s. Dr. Rawat happened to be in California, an hour's drive from me, visiting his son and daughter-in-law. I drove down and had coffee with the family. We had a lovely time, and Dr. Rawat and I agreed to get together in India for a week or two while he investigated whatever case next presented itself.

The Kirti Rawat who met me at the airport was in a less contented state. He had been arguing with management over the room service at the hotel where I had booked us. The next morning, we packed up our bags and moved across Delhi to the Hotel Alka ("The Best Alternative to Luxury"), where he and Stevenson used to stay. The carpets are clammy, and the toilet seat slaps you on the rear as you get up. The elevator is the size of a telephone booth. But Dr. Rawat likes the vegetarian dinners, and the service is attentive to the point of preposterousness. Bellhops in glittery jackets and curly-uppy-toed slippers flank the front doors, opening them wide as we pass, as though we're foreign dignitaries or Paris Hilton on a shopping break.

It is 9 a.m. on the first day of our travels. A driver waits outside. This is less extravagant than it sounds. The car is a 1965 Ambassador with one functioning windshield wiper. Dr. Rawat seems not to mind. The most I could get out of him on the subject of aged Ambassadors was that they are "beginning to be outmoded." What he likes best about this particular car is the driver. "He is submissive," Dr. Rawat says to me as we pull away from the curb. "Generally, I like people who are submissive."

Oh, dear.

This week's case centers on a boy from the village of Chandner, three hours' drive from Delhi. Dr. Rawat is using the drive time to fill me in on the particulars of the case, but I'm finding it hard to pay attention. We are stuck in traffic just outside Delhi. There are no real lanes, just opposing currents of vehicles, chaotic and random, as though they'd been scooped up in a Yahtzee cup and tossed haphazardly onto the asphalt. Every few feet, a cluster of cows has seemingly been Photoshopped into the mix: sauntering mid-lane or lying down in improbably calm, sleepy-eyed pajama parties on the median strips. We enter a lurching, kaleidoscopic roundabout. In the eye of the maelstrom, a traffic cop stands in a concrete gazebo, waving his hand. I cannot tell whether he is directing traffic or merely fanning himself.

I wonder aloud where all these people are going. "Everyone is going to his own destination," comes the reply. This is a highly Dr. Rawat thing to say. One of Rawat's two master's degrees and his doctorate are in philosophy, and it remains one of his passions—along with Indian devotional music and poetry. He is the dreamiest of scientists. Last night, in the midst of a noisy, hot, polluted cab drive, he leaned over and said, "Are you in a mood to hear one of my poems?"

Dr. Rawat is telling me that the case we are investigating is fairly typical. The child, Aishwary, began talking about people from a previous existence when he was around three. Ninety-five percent of the children in Stevenson's cases began talking about a previous existence between the ages of two and four, and started to forget about it all by age five.

"Also typical is the sudden, violent death of the P.P."

"Sorry—the what?"

"The previous personality." The deceased individual thought to be reincarnated. "We say 'P.P.' for short." Possibly they shouldn't.

Aishwary is thought by his family to be the reincarnation of a factory worker named Veerpal, several villages distant, who accidently electrocuted himself not long before Aishwary was born. Dr. Rawat opens his briefcase and takes out an envelope of snapshots from last month, when he began this investigation. "Here is the boy Aishwary at the birthday party of his 'son.'" Aishwary is four in the photograph. His "son" has just turned ten. Just in case the age business isn't sufficiently topsy-turvy, the elastic strap on the "son's" birthday hat has been inexplicably outfitted with a long, white beard. This morning, while leafing through a file of Dr. Rawat's correspondences, I came across a letter that included the line: "I am so glad you were able to marry your daughter." I am reasonably, but not entirely, sure that the correspondent meant "marry off."

"Now, here is the boy with Rani." Rani is the dead factory worker's widow. She is twenty-six years old. In the photo, the boy stares fondly—lustfully, one might almost say, were one to spend too much time in India with reincarnation researchers—at his alleged past-life wife. This strikes me as the most improbable, chimerical thing I've ever seen, and then I look out the car window, where an elephant plods down a busy Delhi motorway.

Living in California, where alleged reincarnations tend to spring from royalty and aristocracy, a reincarnated laborer is something of a novelty for me. Dr. Rawat says that this is typical here: "These are ordinary people remembering ordinary lives." Though there are exceptions. At last count, he has met six bogus Nehru reincarnates and eight wannabe Gandhis.*

---

*Delusions of reincarnation typically fit the culture and religion of the deluded—Saddam Hussein claiming to be Babylonian king Nebuchadrezzar, excommunicated Mormon polygamist cult leader James Harmston claiming to be Joseph Smith, et cetera. Jesus is your big exception. A Google search on "claims to be the reincarnation of Jesus" turns up thirty-one competing candidates for J.C. incarnate, including the Reverend Sun Myung Moon—rumored to add drops of his own blood to the communion wine—and a Mr. Fukunaga, leader of an obscure Japanese foot-reading sect. Mr. Fukunaga also claims to be Buddha reincarnate. This is less impressive than it sounds because—according to the "Jesus Reincarnation Index" of New Age author Kevin Williams—Jesus is a reincarnation of Buddha.

In the case of the boy Aishwary, the alleged previous personality hails from a family just as poor as his own. In Dr. Rawat's estimation, this strengthens the case, as financial gain wouldn't be a motive for fraud. Poor families have been known to fabricate a rebirth story in the hope that the "previous personality's" family—they'll target a wealthy one—will feel financially beholden to their dead relative's new family. Dr. Rawat told me about another creative application of ersatz reincarnation: escaping an unpleasant marriage. Years back, he investigated the case of a woman who fell ill and claimed to have momentarily died—and then been revived with a different soul. Now that she had been reborn as someone new, she argued, she couldn't possibly be expected to live or sleep with her old husband. (Divorce retains a weighty social stigma in India.) Dr. Rawat interviewed the doctor who examined her. "He wasn't a doctor at all. He was a compounder." A bone-setter. And she wasn't dead. "He told me, 'Well, her pulse was down.'"

While Dr. Rawat catnaps, I page through a copy of his book *Reincarnation: How Strong Is the Scientific Evidence?* Let's set aside "strong" for a minute and talk about "scientific." Like most psychological and philosophical theories, reincarnation can't be proved in a lab. You can't see it happen, and no biological framework exists to explain how it might work. The techniques of reincarnation researchers most closely match those of police detectives. It's an exhausting, exacting search for independently verifiable facts. Researchers contact the parents of the child and then travel to the village or town. They ask the parents to recall exactly what happened: word for word, detail by detail, what the child said when he first began speaking about people or places from a past that clearly didn't correspond to the life he now lives. They look for credible witnesses to the child's utterances, and they interview them, too.

By the time the researcher arrives on the scene, the family

has usually found a likely candidate for the child's former incarnation. Most Indian villagers accept reincarnation as fact, and word of a child remembering a past life travels quickly to neighboring villages. The previous personality can't be interviewed, because he's dead, but his family members can. If the child is said to have recognized his home from his past life or features of the town or members of his past-life family, the researcher interviews witnesses who saw the meetings and the purported recognitions.

The strongest cases are those in which the parents have written down the child's statements when he or she first began talking about a past life—before they've met any family or friends from that life. (These are rare: Among Stevenson's cases, only about twenty include any written record.) Without a written record, researchers must work from the parents' memories of what the child said. This makes for wobbly evidence—not because villagers are dishonest, but because human memory is deeply fallible. It's unreliable and easily tweaked by its owner's beliefs and desires. Did the boy say what he said about electrocution before his parents began talking about Veerpal's death, or did he perhaps overhear them talking about it first? Did he really say he was killed by an electrical current, or has his mom, once she learned the facts, reinterpreted something ambiguous? Perhaps the boy referred to a cord. He meant a rope, but the mother, having heard about the accident, pictures an electrical cord. That sort of thing.

Most of Ian Stevenson's case write-ups include a chart summarizing the allegedly reborn child's statements about a past life and about people he or she recognizes. For each of these statements and recognitions, Stevenson lists a witness, if there is one, and the comment of the witness. Typically the chart marches on for eight or ten pages, wearing down your skepticism with the grinding accumulation of names and tiny type. If you take the work of Ian Stevenson at face value, it would be hard to reach any conclusion other than this: Reincarnation happens.

The skeptics tend to dismiss Stevenson's work a priori; few have taken him on case by case. One who tried was Leonard Angel, then a humanities professor at Douglas College in British Columbia. He chose the case of a Druze boy from Lebanon, Imad Elawar, a case Stevenson has referred to as one of his strongest. Of all the cases in which there is written documentation from the time before the suspected previous personality was located, this is the only one in which Stevenson himself wrote the statements down—thus precluding a fraudulent after-the-fact jotting. Angel complains that Stevenson nowhere sets forth these statements as they were worded by the boy or the parents. Stevenson simply writes that the parents "believed [the boy] to have been one Mahmoud Bouhamzy of Khriby who had a wife called Jamilah [Mahmoud and Jamilah were the names the boy spoke first and most often] and who had been fatally injured by a truck after a quarrel with the driver."

Stevenson traveled with the family for their first visit to Khriby. He couldn't find a suitable Mahmoud Bouhamzy; however, upon asking around, he found an Ibrahim Bouhamzy with a mistress named Jamilah. Ibrahim was not run over by a truck, but his relative Said was, though no quarrel was involved. Stevenson concludes that the boy's parents had made wrong inferences based on his words—though since his write-up does not give the boy's exact words, it's hard to know what to think. There's no explanation of why the name most commonly uttered by the boy would be Mahmoud. The glass slipper fit Ibrahim, and Stevenson proceeded from there.

But I was never in Khriby, and neither was Leonard Angel. Something served to convince Stevenson that the case of Imad

Elawar strongly suggested reincarnation. Whether it was the facts of the case or a blind eye born of bias, I can't say.

So I've come to India for answers. I want to get inside one of these cases, meet the families involved, hear the things they say, watch them interact.

In India, I'm finding, the answers do not always fit the questions. This morning at the hotel, I asked the waiter what kind of cheese is in the masala omelette.

"Sliced," he said.

I hope to do better than that.

THE TRAFFIC JAM has dissolved, leaving our driver free to proceed in the manner he enjoys. This entails driving as fast as possible until the rear end of the car in front is practically in his mouth, then laying on the horn until the car pulls into the other lane. If the other car won't move over, he veers into the path of oncoming traffic—for sheer drama, an approaching semi truck is best—and then back, at the last possible instant. Livestock and crater-sized potholes materialize out of nowhere, prompting sudden James-Bond-style swervings and brakings. It's like living inside a video game.

"Why doesn't he just get into the fast lane and stay there?"

"There isn't a fast lane, as such," says Dr. Rawat. He gazes calmly out his window, as goats and a billboard for Relaxo footwear flash past. "The lanes are both the same. Whoever is slower pulls over." He speaks in a neutral, narrative tone, as though describing a safe and civilized code of the road. Aggressive honking and light-flashing is considered good manners: You're simply alerting the driver ahead of your presence. (Rearview mirrors are apparently for checking your hairdo. Likewise, the driver's-side mirror currently registers a clear and unobstructed view of the dashboard.) Exhortations to BLOW HORN PLEASE and USE DIPPER are painted on the backs of most trucks, so that even the most laid-back driver goes along honking and flashing his lights like his team has just won the World Cup. I am finding it hard to relaxo.

In India, everywhere you look, people are calmly comporting themselves in a manner that we in the States would consider a terrible risk, a beseeching of death with signal flare and megaphone. Women in saris perch sidesaddle, unhelmeted, on the backs of freeway-fast Vespas. Bicyclists weave through clots of city traffic, breathing diesel fumes. Passengers sit atop truck cabs and hang off the sides like those acrobat troupes that pile onto a single bicycle. Trucks overladen with bulbous muffin-top loads threaten to topple and bury nearby motorists under illegal tonnages of cauliflower and potatoes. (ACCIDENT PRONE AREA, the signs say, as though the area itself were somehow responsible for the carnage.) People don't seem to approach life with the same terrified, risk-aversive tenacity that we do. I'm beginning to understand why, religious doctrine aside, the concept of reincarnation might be so popular here. Rural India seems like a place where life is taken away too easily—accidents, childhood diseases, poverty, murder. If you'll be back for another go, why get too worked up about the leaving?

A bus blasts its horn and bullies us onto the shoulder. "&⋆@##!!"

Dr. Rawat winces. "Meddy! Just don't look out that side!"

We've been bickering all morning. Dr. Rawat let it be known that he booked me for three appearances in his home city, including a talk on the theme of "teacher appreciation" at the Indore Lion's Club. He has me in Indore for four days, when I had planned on two. I tried to use the excuse that I have nothing to wear. He suggested I wear one of his wife's saris. "The sari," he said when I balked, "is the *most* elegant dress for women." At one point he said, "You do not dress to please yourself; you dress to please others." You can imagine how well that went over. Poor Kirti. He wanted vanilla and he got jalapeño.

Today's plan is to head first to Chandner for some follow-up interviews with Aishwary's mother, and then drive, along with Aishwary's family, to two neighboring villages where the family of Veerpal, the boy's alleged previous personality, resides.

As we approach Chandner, Dr. Rawat summarizes the family's claims. The boy's father, Munni, claims that Aishwary recognized Veerpal's uncles and aunts when they came to Chandner, and that he could name many of the people in one of Veerpal's photo albums. He further claims that the boy said he had three children and family members living in Kamalpur, and that his caste was Lodh, all of which are true of Veerpal. When Munni went to buy a sari as a gift for Veerpal's widow Rani, Aishwary is said to have insisted that it be turquoise. Veerpal, Rani says, used to buy her saris in this color. Munni further reports that Aishwary was spotted hitting an electrical pole with a stick and calling it "abusive names." Munni's wife Ramvati says she saw Aishwary try to kiss Rani on the lips and that the boy was spotted caressing her breast.

Dr. Rawat says this sort of sexual precociousness is an infrequent but not unheard-of by-product of rebirth cases. "That is nothing. I heard of a case where a husband said to his wife, 'When I die, I will come back as your son, and I won't take milk from your breast.'" Sure enough, the story goes, the husband died during his wife's pregnancy, and the infant born some months later refused to breast-feed. "It is said she was both his mother *and* his wife."

"That's what all you men want," I say. "Not that there's anything wrong with it."

Aishwary's family grows corn and sugarcane. As we walk through their rain-boggy yard, we pass the kernels of this season's harvest spread over the concrete floor of the house to dry. A pair of oxen lounge in the mud. Their horns spiral like curling ribbon on the sides of their heads. Up a flight of outdoor stairs and across a rooftop is the family's single-room sleeping quarters. The room holds little aside from three caned, wooden sleeping platforms and a flickery black and white TV.

Aishwary's mother boils water for tea, squatting over a hot plate in the corner. Dr. Rawat sits on a bed beside Aishwary and shows him the photos from the birthday party last month. He points to the boy with the strap-on beard. "Who's this?" He translates the boy's reply: "This is my son." Some of the other pictures are met with blank looks. Even when handed a picture of the electrocuted Veerpal, he shakes his head and looks toward his mother. "He doesn't seem to remember much now," says Dr. Rawat.

Aishwary's father Munni fills us in on new developments in the case. Like his wife, Munni has a sunny smile and a pleasing, well-proportioned face. He is telling Dr. Rawat that Aishwary walked up to a boy in Veerpal's town and said to him, "Your parents came to see me in the hospital." The parents confirmed that they had gone to see Veerpal after the accident. Munni adds that Veerpal's aunt, while clowning around with Aishwary, reported that the boy said to her, "Auntie, you have not left your old habits," and that this was said to be the exact wording of a phrase Veerpal used to use. Dr. Rawat makes a note of this, as we'll be visiting the aunt this afternoon.

Before leaving for the aunt's village, we walk across the town to visit another boy who is said to recall a past life. Indian villages are fertile ground for claims of reincarnation. "You come for one," says Dr. Rawat, "and you leave with four!"

This cannot be said of villages or cities where reincarnation isn't part of the belief system. Claims of reincarnation are rare among children in the United States, where—according to a 2001 Gallup poll—only twenty-five percent of the population believes in it. This fact, perhaps more than any other, weakens the overall case for reincarnation. Stories of rebirth that crop up within cultures whose religious dogma doesn't include it are, for obvious reasons, stronger than cases that show up among cultures who accept it and, more to the point, expect it to happen. If a child in a Western culture begins to refer to a stranger with an unfamiliar name, his parents assume the name belongs to someone from his imagination. In a Hindu—or Druze, or Tlingit—culture, the parents are more likely to assume it's someone from his past life. Are cases solved, or are they built? "This is the most common criticism of reincarnation research," says Jim Tucker, professor of psychiatric medicine at the University of Virginia, who researches cases in the United States. Stevenson agrees. "I don't have a good explanation for that," he told an *Inside UVA* interviewer. "I worry about it." Stevenson and Rawat suggest that the difference may arise from the parents' reactions: In a culture that embraces reincarnation, the child may be encouraged to voice his memories; anywhere else, the child's comments may be ignored—or thought abnormal and thus discouraged.

Dr. Rawat is excited about the new case in Chandner, as it's a Hindu boy who recalls a past life as a Muslim. (More exciting, for the reasons just given, would be a Muslim boy who recalls being Hindu.) A crowd has formed in our wake. Many are children. We seem to pull them out of houses as we pass. You get the feeling there isn't much for kids to do here. On our way in, we drove past a boy with a paper kite. There was no wind; he merely swung it in circles on its string. We're the most exciting thing to hit town since electricity.

Dr. Rawat is telling me about another Muslim-to-Hindu case from some years back. "He remembered the process of circumcision," he says to me, picking his way from brick to brick through the muddied street. "And moreover! He was born with a penis without a foreskin!"

I was about to ask Dr. Rawat whether he thinks that the unique circumstances of the penis may have inspired the boy's imagination and/or the parents', but my flip-flop has been claimed by the sucking mud. When I pull on it, the rubber shoe slingshots out of its sinkhole and spatters the back of my skirt. Boys and girls titter and squeal: Why, this is as good as it gets!

As we arrive at the boy's house, our following has grown to fifty or more. Dr. Rawat doesn't like to do interviews in front of a crowd, lest the subject feel pressured to answer one way over another. He closes and bars a corrugated tin gate. The crowd presses in. The panels bang and bow and threaten to give, like a boudoir door in a cheap suspense film. We sit down on a porch to talk to the grandparents of the alleged former Muslim. (The parents are away.) Onlookers have scaled the buildings across the street. They squat at the roof's edge and peer down at us like gangly, brown-eyed gargoyles. On the wall, a single shelf is lined with a sheet of newspaper scissored to resemble the zigzag-fringed doilies of middle-class homes such as Dr. Rawat's. "Four Cheers!" says a headline in a digital camera ad. "The Future Has Come Calling!"

The boy, who is seven, claims to recall a life as a Muslim thief named Guddin in the town of Dhampur, seventy kilometers away. Dr. Rawat translates for me. "I killed two policemen, and then they killed me." Discussion ensues. Laundry drips on my head. "Someone else says twelve policemen," Dr. Rawat narrates. "The grandparents add that the boy has always had a fear of police cars. The boy said his wife was Dhamyanta, but that's not a Muslim name. Come, we shall have some photographs of his penis." He wants to see whether perhaps this child, too, has a birth defect that mimics circumcision. "We will verify his foreskin."

Dr. Rawat, myself, the boy, and the boy's grandfather slip into the house and close the door. The grandfather picks up the boy and stands him on a table. The boy unfastens his shorts and turns his face away from us. He doesn't seem upset by the request, just embarrassed. His foreskin is normal, but Dr. Rawat aims the camera anyway. It's a new one that he's not yet accustomed to. Seconds pass, as though he's waiting for the tiny member to smile. I point to a button on the back. A red light comes on. Oh, good. We've activated the anti-red-eye function. If ever there were a moment that wanted to pass quickly, this is it. At last the flash goes off and the boy is free to cover up.

A few words about birth defects and birthmarks. Among cultures that believe in reincarnation, congenital abnormalities are commonly viewed as clues to a child's past life. Often they are tied in with the death of the supposed previous personality. Ian Stevenson's *Reincarnation and Biology* contains ten examples of children with birthmarks or birth defects corresponding to the place their alleged previous personality was shot or otherwise fatally wounded.

The birthmark business has a historical corollary of sorts in the theory of maternal impressions. A surprising majority of sixteenth- and seventeenth-century physicians believed that a child's birthmarks or abnormalities are caused by the mother having undergone a memorable fright during pregnancy. A baby is born with a missing arm; the mother recalls being set upon by a one-armed beggar. A child's "fish scales"—a skin condition now known as ichthyosis—are blamed on the mother's fear of sea serpents. Et cetera.* Reports of maternal impressions peppered medical texts from Pliny and Hippocrates clear through to the 1903 edition of the *American Textbook of Obstetrics*, which cites maternal impression as the likely cause of John "Elephant Man" Merrick's deformities—as well as those of a lesser-known traveling spectacle, the Turtle Man.

In many of the birthmark cases in *Reincarnation and Biology*, Stevenson posits that the mother saw the corpse of the slain man whose soul eventually turns up in her unborn child. Stevenson doesn't believe all birthmarks are caused by maternal impression, but he is open-minded to the possibility that some are.

Adherents of maternal impression theory hold that the skin is uniquely vulnerable to emotional imprinting. Stevenson describes a half dozen dermatological conditions thought to be open to psychological influence. These range from the relatively mainstream (emotionally induced wheals and blisters) to the distant borderlands of scientific acceptability (stigmata, wart-charming, hypnotically induced breast enlargement). I suppose that if you believe that hypnotic suggestion can expand a bosom, it's not a big leap to suppose that

---

*By some accounts, Mom didn't need to be frightened but merely focused a little too long in one place. In a famous case detailed by Jan Bondeson in *A Cabinet of Medical Curiosities*, a thirteenth-century Roman noblewoman gives birth to a boy with fur and claws; the authorities lay blame on an oil painting of a bear on her bedroom wall. The event prompted Pope Martin IV, clearly a tad hysterical, to have all pictures and statues of bears destroyed.

Crafty moms tried to work the phenomenon in their favor. In the early 1800s, Bondeson writes, it was common for pregnant noblewomen to be wheeled into the Louvre to spend an hour or so gazing at a portrait of some handsome earl or archduke of yore, in hopes of influencing their unborn progeny.

a profound fright might affect the skin of a developing fetus.

What of the boy with the missing foreskin? Was his previous personality's penis the site of a fatal injury? Unlikely. This is more a case of a suggestive similarity. Stevenson and the families he talks to also make connections based on simple physical and psychological parallels between a child and the person they believe he or she once was. Stevenson feels that genetics and environmental influences fall short of adequately explaining the quirks and foibles—both medical and psychological—that we humans are born with. He looks to the quirks and foibles of the previous personality to explain what genetics cannot. The concept has a certain intuitive appeal. A child's former life as a World War II soldier explains a fear of Japanese people. A past life as a virtuoso musician explains a musical prodigy in a family of tone-deaf no-talents. Yet you've simply swapped one mystery for another. How—outside of genetics—would the dead person's skills, fears, or preferences be delivered to the new organism? What's the mechanism? Here we don't even have the flimsy leg of maternal impression to stand on.

Unconstrained by biology, Stevenson is free to extend his theory wherever it strikes him. Facets of a past life are suggested as explanations for complexion irregularities, stockiness, third nipples, albinism, posture, gait, fear of women, fondness for toy airplanes, cleft lip, pimples, speech impediments, widely separated upper medial incisors, and "a fondness for eels, cheroots and alcohol." Viewed through such a broad eyepiece, reincarnation is an easy sell. Take a child and all her hundreds of unique features: How hard would it be to find one or two that seem linked to a feature of someone you know who has died?

The notion is especially rickety when you consider that in many of Stevenson's cases, the life recalled by the child is that of a close blood relative. Why posit reincarnation when you've got a perfectly reasonable biological explanation in the form of genetics? Even Ian Stevenson's wife appears to have trouble swallowing the whole bolus. In his acknowledgments in *Reincarnation and Biology* he writes, "While devotedly encouraging this work, she has also—with the greatest gentleness—expressed skepticism about the conclusions to which it has led me."

We have walked back to Aishwary's house to pick up the family for the trip to visit Veerpal's aunt in Kamalpur. Aishwary is changing his clothes for the visit. Dr. Rawat, ever vigilant for birthmarks and scars, bends to inspect a semicircular protrusion in the middle of the boy's chest.

"Do you think this is anything?" he asks me.

"I think it's a sternum."

WE REACH KAMALPUR just after 2 p.m. Word spreads instantly. The boy is here! The future has come calling! A crowd surrounds the car well before the driver has cut the motor. "Faster than flies to sweets!" exclaims Dr. Rawat. Or flies to more or less anything. The moment we stop moving, logy black ones alight on my arms, my skirt, the upholstery beside me. The situation is not helped by the fabric's pattern, which is little bees.

We get out and begin walking to the house of Veerpal's aunt Sharbati. Many of the women in the streets have draped their sari scarves over their faces in modesty; curiously—to me, anyway—their midriffs are partly bared.

Dr. Rawat stops the procession beside a tree with a small shrine beneath it. Munni said that his son had talked about there being a shrine behind his aunt's house. This is the shrine he is said to have recognized. "And there"—Dr. Rawat turns 180 degrees and points to a faded blue doorway down a street, about a half a block distant—"is the house." So it's behind the front of the house. In other words, it's no more behind this house than any other house within eyesight.

The sameness of the villages in this part of India renders less impressive some of the children's statements in these cases. "The floor was of stone slabs." "The family had cows and oxen." "The house had two rooms." Facts like these could apply to a dozen houses in any given village. Still, the casebooks are full of true statements so specific that—if in fact the child made them, and if the family had never visited the past-life town—defy logical explanation: "He had a wooden elephant, a toy of Lord Krishna, and a ball on an elastic string." "He had a small yellow car." It's hard to know what to make of it.

Veerpal's aunt traveled to Aishwary's village several weeks ago, but this is the first time Dr. Rawat has met her. The house is of the standard two-room floor plan. Like most houses here, the front room has three walls only. As we walk by, domestic scenes are on display like shoebox dioramas. A toddler plays with a corncob, making believe it's a cigar. A woman stacks dried ox dung. A man gets a shave.

Dr. Rawat begins rolling video of the aunt, despite the crowd. When there are no doors, there is little to be done about it. I count fifty-five pairs of feet gathered around, most of them shoeless. Kamalpur is even poorer than Chandner. My glance takes in broken trouser flies and saris patched with duct tape. Here again, Dr. Rawat is encouraged: Skeptics often cite monetary motives for making up claims of rebirth. In Aishwary's case, the family of the current incarnation has given about as many presents to the past family (saris for the widow) as the past family has given to the boy (hundred-rupee notes tucked in his pockets).

The press of the crowd has created its own weather system, a thick, clinging humidity that lies on your skin like glaze.

Aishwary yawns and drops his head in his mother's lap. Veerpal's aunt has a smoky voice and one turned-in eye, and she stands with one hand on a jutted hip. Overall she strikes me as someone you'd go out of your way not to cross. Dr. Rawat tells her over and over to relate only the events and statements that she herself has seen or heard the boy say. He asks her about the line Munni mentioned: "Auntie, you have not left your old habits." She says that the boy indeed said this, but the bit about Veerpal having used this phrasing is not true. The utterance suggests only that the boy believes himself to have been reincarnated as Veerpal, which is not, given the culture and the fact that his parents clearly believe it, all that surprising.

More difficult to explain is the account of Veerpal's uncle Gajraj, whom we visit next. He is a schoolteacher in the village, a somber, balding man dressed in white dhoti pants and tunic. "Tell me what you saw and heard," says Dr. Rawat as tea and sweets are served in the front room of his two-room home. Above the doorway, a pair of old wood badminton racquets is mounted like crossed swords in a coat of arms. A young boy stands by my side, fanning us with a stiff, laminated flag.

"I was returning from my farm," begins Gajraj, "and as I entered the village, people said, 'Veerpal has come!' I was astonished. How could Veerpal come? There were two or three hundred people. The child said nothing at that time. Then Mokesh was called." Mokesh was a close friend of Veerpal's. "The headman of the village arrived and asked the boy if he recognized Mokesh. The child said nothing. The headman said, 'What is his name? Say it in my ear.' And he did. We could hear him say, 'Mokesh.'"

"You heard it yourself?"

"Yes."

"What else?"

"He pointed toward me and said, 'You are my uncle.'"

"Did he say your name?"

"No."

He adds that the boy recognized Veerpal's sister. "He said, 'She is my sister, Bala.'" The deadpan and monotone of Gajraj and Veerpal's other family members are puzzling to me. These conversations and encounters with Aishwary hold no more emotion than a market research interview about soap-buying habits. The only animation in the small room comes from the fan boy, who is waving vigorous, exaggerated figure eights. (I'm still hot, but I feel like I've won the Indy 500.) If I'd lost my brother or my nephew and then, months later, come to believe that he'd been reborn as a boy in a neighboring village, it would be a story I'd tell with feeling and awe. Perhaps the video camera makes them self-conscious. And, to be fair, I'm not witnessing a first encounter with the boy. That will come at our next stop, the village of Bulandshahar, where Aishwary will meet Veerpal's father for the first time.

Toward the end of his interview, Gajraj is asked whether he believes that this boy is his nephew reborn. He says yes, and adds that it is not the first reincarnation he has encountered. "In my classroom, I recognize many children again and again."

Gajraj's two brothers, whom we next interview, seem less convinced of the boy's status as their reborn nephew. Both report that the boy did not recognize them.

"What do you think?" Dr. Rawat asks the third uncle at the end of the interview. "Do you believe that this boy was Veerpal?" The uncle, dressed in a white singlet and a layer of perspiration, looks uncomfortable. "I can't say."

TO SAY THAT HINDUS believe in reincarnation is in and of itself rather meaningless. Catholics "believe" that they are eating the body of Christ when they take communion, but how many believe it literally?* I used to assume that people in India believed in reincarnation in the same way that Christians believe in heaven: more or less abstractly. Most Christians don't expect to take up residence in a cloud bank after they die, but they may believe in an abstract sense of the hereafter as a place whose comforts or lack thereof depend upon one's behavior here on earth.

I began to change my tune after spending an afternoon among the pages of *The Ordinances of Manu*, a tome of legal code based on Vedic scripture and dating back to A.D. 500. Manu's legislation covers everything from criminal law (If a man of the lowest birth spit upon a highborn man, "the king should cause his two lips to be cut off; and if he make water upon him, his penis; and if he break wind upon him, his buttocks") to health and hygiene codes ("Anything pecked by birds, smelt by a cow, . . . sneezed on or polluted by head lice becomes pure by throwing earth on it")—and reincarnation is in there, too.

In Manu's day, reincarnation was treated not as an abstract religious principle but as a concrete legal consequence. Where the modern-day malefactor may do time in Pelican Bay, the

*And are we meant to? Unsure despite my Catholic upbringing, I consulted *The Celebration of Mass*, as thorough a manual of Catholic ritual as you'll find outside the Vatican. While nowhere was it stated that the consecrated host is literally Christ, it is most certainly treated as something beyond a quarter ounce of unleavened wheat flour. For example, one may not simply throw old, stale hosts into the garbage; they must be consumed by the priest, unless they are so moldy as to be inedible, in which case they are to be burned or mysteriously "done away with in the sacrarium." And finally, "should anyone vomit the Blessed Eucharist, the matter is to be gathered up and put in some becoming place."

perpetrator in Manu's day might do time as an actual pelican. Witness Code 66 of Chapter XII: "One becomes indeed a kind of heron by stealing fire; a house-wasp by stealing a house utensil; by stealing dyed cloths one is born again as a fowl called *jivijivaka*." Similarly, for stealing silk, linen, cotton, a cow, or molasses, one is reborn, respectively, as a partridge, a frog, a curlew, an iguana, or a *vagguda* bird. The worst karmic punishments are reserved for those who "violate the guru's couch." I am unclear on precisely what is meant by this, but my guess is that we are not speaking of a literal rending of upholstery, for the hapless malfeasant is sentenced to return "hundreds of times into the womb of grasses, bushes, vines, animals that eat raw flesh, . . . and animals that have done cruel acts." Similarly unwise is the Brahman who has "deserted his own proper rules of right," for he must reincarnate as "the ghost Ulkamukha, an eater of vomit."

The point I was trying to make, when I became helplessly distracted by the quixotic deemings of Manu, is that reincarnation has traditionally been accepted as a literal, not allegorical, facet of life. The villagers I am meeting this week do not question whether the dead are reborn, any more than we would question whether they decompose. Veerpal had to enter someone else, why not Aishwary? I'm not saying the events of these cases are untrue; I'm saying that no villager is likely to judge them with an especially critical eye or ear. And, also, that "one should not voluntarily stand near used unguents" (Chapter IV, Code 132).

THE ROAD TO Bulandshahar, the home of Veerpal's parents, takes us through a sprawling outdoor marketplace. Reincarnation is going on all over the place: eight old Vespa hulls rest on the dirt outside a mechanic's shed, awaiting new engines. Shoes are resoled, electric fans gutted and reworked. A boy pushes a filthy rusted bicycle, seat worn down to its metal skull, to the stall of a tire vendor, where rims hang like bangles on a rope between two trees. Aside from fruit and packets of pan and one array of surreally pristine porcelain squatter toilets, nothing for sale here is new. Exteriors are endlessly replaced, and the core carries on.

Veerpal's parents live twenty miles from Kamalpur, and Aishwary's parents used to live nearby. "Scientifically, the proximity of the two families is a weak point," Dr. Rawat is saying. A child who is said to know things about a family of far-off strangers makes a stronger case for reincarnation than a child who is said to know things about a family in a town his parents know well. Weakest—and quite common—are the cases in which the child seems to be the reincarnation of one of his own family members. Stevenson's casebooks hold many of these. In the cultures that most often report it, within-family reincarnation is expected. It's what happens when you die. Among rural Indians, the soul often wanders farther afield, but rarely much beyond a hundred miles.

I ask Dr. Rawat why the human spirit is such a homebody. From what I've been given to understand about the speed and ease of "astral" travel, you'd think a soul might be impelled to hop a continent every now and again. Dr. Rawat shrugs. "You are more comfortable in your own surroundings. You fit in well again." I guess he's got a point.

I had wanted to see Aishwary's face as he casts his first glance at the man believed to have been his father, Mathan Singh. Somehow I fell behind the crowd and missed the moment. So did Dr. Rawat. We step into the room just as

Aishwary is settling into the man's lap. Mr. Singh has a sweet, deeply lined face. He is shy, and so thin you can see the shape of his knee bones pressed together underneath his tunic.

"See how the boy comes into his lap?"

"Kirti, he picked him up and *put* him in his lap."

"See how comfortable the boy looks?"

"He looks just like he did when I held him in my lap yesterday. He's a comfortable boy."

I'm working myself up to full nitpicker skeptic mode, but then something happens. I've been watching Mathan Singh, wondering why he isn't staring deeply into the boy's eyes to try to figure out if it's true, trying to connect with the soul of his lost son somehow. I guess I'd been expecting a Demi-Moore-in-*Ghost* kind of moment, the part where she somehow senses that (God help her) her dead husband is there inside Whoopi Goldberg. What I notice instead is that Mathan Singh, sitting chatting with his arms around the boy, looks profoundly content. It occurs to me that it doesn't much matter whether this boy does or does not hold the soul of the son Mathan Singh lost. If Mathan Singh believes it, and if believing it eases the grief he feels, then this is what matters. It also occurs to me that I don't speak Hindi, and that I have no idea what this man is saying or feeling or believing. He could be saying, "This reincarnation crap. I've never bought it."

I tug on Dr. Rawat's sleeve. "Can you ask him how he feels about all this?"

Dr. Rawat obliges. "He says he is happy. He says, 'My son is alive, therefore I am happy.'" Past-life therapy.

Meanwhile, out the back door, Aishwary's two mothers are laughing together and drinking tea. I might have thought there'd be jealousies and rivalries between the mothers, but Dr. Rawat says he has rarely seen this. It's all a happy excuse for a party.

Since Aishwary's (current) mother and his "wife" met, they have gotten together five times, including one three-week visit.

A group of young men in Western dress has just arrived on the scene. One introduces himself. He is Nathan, visiting from Delhi. City dwellers in India are much less likely to believe in reincarnation, and I ask him what he thinks.

Nathan looks around the room. "Marvelous, ma'am!"

MY FIRST DAY on the streets of Delhi, a live rat dropped from somewhere overhead. It was not thrown, for it descended in a vertical path directly in front of my face, landing more or less on my shoe. It appeared to have simply lost its footing at the precise moment that fate had arranged for my arrival there on the sidewalk. The event struck me as an appalling close call, a brush with vileness and possible scalp laceration, a harbinger of coming horrors and shortcomings in public hygiene.

"Oh!" exclaimed Dr. Rawat. He was as surprised as I was, but here our reactions parted company. "You are blessed! The rat is the conveyance of Lord Ganesha!"

The episode got me thinking. If you are enough of a Hindu to view a falling rat as an auspicious event, are you too much of a Hindu to dismiss reincarnation—if indeed that is what the facts suggest you should do? I wondered about Dr. Rawat's capacity for objectivity. He refers to his research as an obsession, an addiction. "Like a drunkard to his bottle, I am to my cases!" he told me when we first met. But is he investigating reincarnation, or merely hunting for evidence in its favor? How can he remain unbiased?

I am about to ask him just this question. We are at an outdoor reception for the launch of a friend's new reincarnation

TV show, in which the main character is repeatedly, energetically murdered by an ever-varying cast of fiends and jealous lovers—affording her ample opportunities for rebirth. I am slated to perform the inaugural clap of the clapboard. (I am, yes, dressed in a sari.) Now they're shooting the opening credits. The director cues a recorded voice-over of booming, hyper-enunciated English: *AS MAN, DISCARDING WORN-OUT CLOTHES, TAKES OTHER NEE-EW ONES, LIKEWISE THE DISEMBODIED SOUL, CASTING OFF WORN-OUT BODIES, ENTERS INTO OTHERS WHICH ARE NEE-EW. . . .*

"As a Hindu," I begin, "you believe in reincarnation. Is it difficult for you, as a researcher, to maintain your objectivity?"

"I am born into a family that believes in reincarnation," Dr. Rawat allows. "And moreover, in my family there was said to be a case of reincarnation. I am aware that there may be some conscious or unconscious bias in me." He insists that this has made him more cautious, rather than less so. "So that my personal belief, my personal experiences, may not infringe on my scientific pursuit, I assume the role of a critic when I study these cases, not a believer."

Dr. Rawat insists he does not fully accept the doctrines of Hinduism. "I believe in all religions and none," he says to me, picking through a plate of vegetable pakora. He finds meaning and guidance in all of them, and also things to reject. He waxes curmudgeonly over Hinduism's never-ending list of required rites and devotions. "Bathe in a particular river and think all your sins are absolved just by taking a bath. This is absolutely nonsense. If you are doing good to others, you are a most religious person."

Our conversation is interrupted by a breathless girl holding out an autograph book. "Ma'am, I have enjoyed *all* of your films!" Earlier, a man asked me what it felt like to meet

President Bush. Apparently the producer sent out a press release full of extravagant misstatements about my career.

Dr. Rawat credits his father for his mistrust of religious dogma. "He taught us that one should not believe a thing merely because it is written in the scriptures." As a student, the young Kirti was drawn to philosophy, but was pushed by his father toward medicine. Parapsychology was the compromise scenario.

I trust Dr. Rawat not to overstate the facts of his cases. And I don't believe that the people he interviewed today were making things up. Does that mean I believe the reincarnation of Veerpal Singh actually happened? Not as such.

I'll tell you what I think might be happening. Over and over, Dr. Rawat would stop his interviewees and counsel them to relate only what they themselves saw or heard. He admits it's almost impossible. Add to that the likelihood that the stories the villagers have heard are inevitably embellished along the way. It's one big heady game of Indian telephone, the same sort of game that turned me into a film star who hobnobs with President Bush. No one sets out to lie, but the truth gets nicked and misshapen.

In the case of Aishwary, Dr. Rawat agrees with me. "There are some disturbing discrepancies," he says, coaxing chickpeas onto a tear of naan bread. "Some of the facts Aishwary might not have recalled as Munni reported him to have recalled. Or, even if Aishwary reported them, he might have picked them up from hearing his father talking to his mother. These are some of the very important pitfalls."

I ask him his overall opinion of the case. He presses his napkin to his lips and sits back in his chair. "My considered opinion about this case is that it is not a strong case at all." He cites the proximity of the three villages. "They are so near to each other that we never know how many informations travel normally"—as opposed to paranormally—"from one to another. Particularly through the father."

Munni's enthusiasm undermines the case. "This is a very, very minus point, a very strongly minus point if your main people are so enthusiastic to find that the case is true." And very often, they are. The villagers Dr. Rawat works with are inclined to view vague or ambiguous statements as evidence. As he puts it, "They will pounce on anything!"

That evening, Kirti and his wife and two of the TV producer's children drive me to the train station. They present me with gifts and big bags of puffed Indian snack foods for the journey. Kirti and his wife lay garlands of marigolds around my neck as though I'm a deity and not the petulant ingrate they've been dealing with all week. I hug Kirti, pressing the flowers so hard they leave stains on our shirts. "I'm sorry about . . . I don't know. I'm not very submissive."

"It's okay. You only lost your mind twice."

YOU DON'T HAVE to be a poorly educated villager to get caught up in a story like Aishwary's and lose your rational rudder. I experienced a similar phenomenon about ten years ago in rural Ireland. I was hitchhiking through County Wexford, where the name Colfer is a common one. My grandmother was a Colfer, and I was keen to sniff out my Irish roots. One day I spotted a butcher shop with a sign over the window: COLFER MEATS. I walked in and asked the butcher, "Are you a Colfer?"

"I am," he said. Three hours later, I was sitting in a pub with nine Colfers and a copy of my family tree spread out between the pint glasses. Some of the first names overlapped, as Irish names will: Catherines and Johns and Margarets.

There was even a Margaret who had emigrated to Chicago—and my father had stayed with an Aunt Margaret in Chicago when he first came to the States.

I clearly recall sensing that the facts didn't all fit, but the feeling faded as the excitement built and the beer flowed. Come closing time, I was hugging my long-lost uncle Mick and promising to keep in touch. New relatives are a novelty and a charm. It's a buzz, and you want to give in to it.

Six weeks later, back at home, my grandmother's birth certificate arrived from the Dublin General Register Office. Her birth date was about ten years earlier than I'd thought. My Irish "family" were no more than friendly strangers in a pub. I'd been swept up in the excitement of the unraveling, paying attention to the facts and dates that fit, overlooking those that didn't.

It is certainly possible that in among the reborn Veerpals and the long-lost Uncle Micks are true links and souls that have lived before. For those with the patience to wade through Ian Stevenson's colossal compilations of case studies, there is much that leaves you scratching your head: statements too specific to suggest coincidence, and no obvious motive for a hoax.

Of late, I find myself wondering about the mechanics of it, the unfathomable blending of metaphysics and embryology. How would the suddenly homeless soul get itself situated someplace new? How does spirit, for want of a more precise word, infuse itself into a clump of cells quietly multiplying on a uterus wall? How do you get *in* there?

Scientists and philosophers of bygone years had a name for the impossible moment. They called it ensoulment, and they debated it for centuries. If the National Science Foundation had existed in the 1600s, there would have been an entire lavishly funded Institute of Ensoulment, devoted to studying the mysteries of human generation, the how and when of life's initial spark. Most of the research covered in this book focuses on things that happen to a person as, and after, his body reaches the finish line, but it makes sense to spend a little time at the other end of the footrace, too.

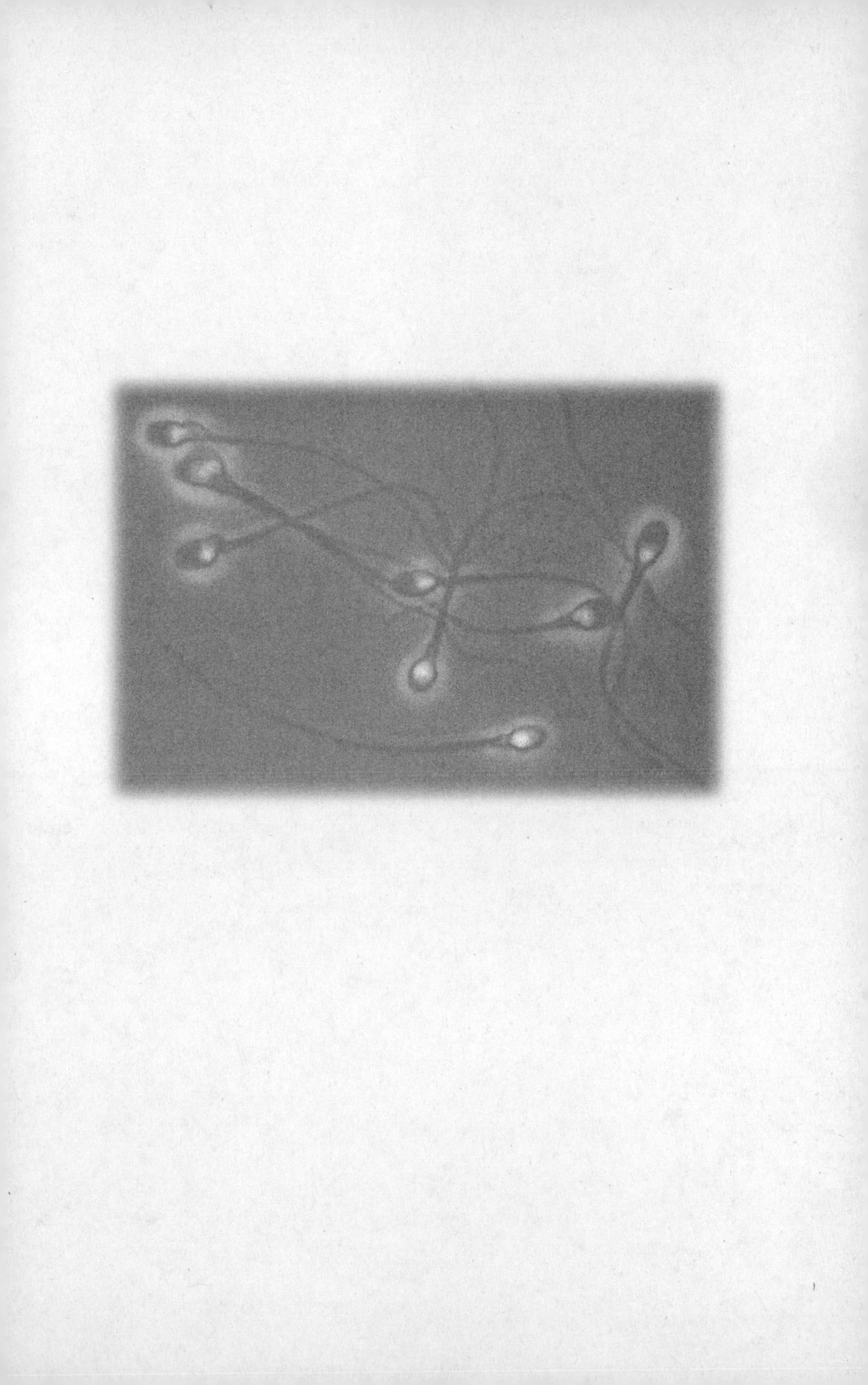

The Little Man Inside the Sperm, or Possibly the Big Toe

Hunting the soul with microscopes and scalpels

THERE'S A VERY good chance that you underestimate the historic import of the sea urchin.* In 1875, a German biologist named Oskar Hertwig watched a sea urchin sperm nose its way into a sea urchin egg

---

*There's a good chance you underestimate almost everything about the sea urchin. For instance, the *Encyclopædia Britannica* tells us some sea urchins use their little sucker-tipped feet to hold pieces of seaweed over their heads like parasols, for shade. Plus, they have teeth that can drill into rock and excavate entire living rooms for their owners. The teeth are hard to see, because sea urchins sit on their mouths; possibly they are self-conscious about their "complex dental apparatus called Aristotle's lantern." One type has spines that can be used as pencils, though not, disappointingly, by the urchin itself.

and fuse with it to form a single nucleus, right there under his microscope. It took civilized man six thousand years to figure out how life begins, and the honors go to humble Oskar and his amorous echinoderms.

Scientists had long suspected that human generation had something to do with eggs—most everyone who owned a chicken suspected this—and they knew it had to do with intercourse and semen, but beyond that they were unclear on the specifics. This was largely because they couldn't *see* the specifics. The sea urchin's eggs offered two advantages: (a) they're see-through, and (b) they're fertilized outside of the female's body, in the ocean or, in occasional cases, under some German dude's microscope.

This means that for six thousand years, there was lots and lots of entertaining speculation about the creation of new human beings. Some of the earliest and most thorough speculating was done by Aristotle. The learned Greek—who, I was interested to read, went through life with a lisp—decided that the man's semen supplied the soul of the new individual. The spirit, in those days, was envisioned as a kind of vapor or breath, which was understandable given breathing's obvious connection to being alive. Hence Aristotle's name for the spirit: *pneuma*, which is Greek for "wind." He believed it was this pneuma, carried in the semen, that orchestrated the creation of a budding human being. Upon arrival inside the uterus, the pneuma would set to work, building new life out of the materials it had on hand: menstrual blood, to be unpleasantly specific. Aristotle described the process as a sort of coagulation, using the apt if inelegant analogy of a cheesemaker's rennet solidifying milk. It took seven days for the new entity to "set," at which time the pneuma would infuse it with the first of three eventual souls. This vegetative soul, as it was called, was a sort of starter soul, a learner's permit for human existence. You were a thing that eats and grows: more than a potato but less than a human.

On the fortieth day, Aristotle theorized, the proto-human morphed into what he called the sensitive soul. By "sensitive," he meant "relating to the senses," for forty days is about when the embryo's sense organs begin to appear. After some further, unspecified amount of time had passed, the pneuma would allow the newly minted sensitive soul to upgrade to a rational soul. Here was the black belt of humanness, the sort of spirit that rises above animal lusts and girly emotions and pays no heed to people who make fun of the way it says *semen*.

And that's pretty much what people believed for the two thousand years after Aristotle put the word out. The man who elevated the ovum to a leadership role in the proceedings was seventeenth-century English physician William Harvey. Harvey is best known for figuring out that blood circulates in a closed system of arteries and veins, a feat he managed by dissecting cadavers, including that of his sister. For his pioneering work in reproduction, you will be relieved to learn that Harvey left the womenfolk alone. Here he turned to a herd of deer that wandered, ever more warily, the grounds of his employer, King Charles I. As a student of Aristotle's teachings, Harvey expected to find the requisite coagulated blob when he dissected the deer's uteruses. He was astonished to instead find the beginnings of tiny deer: embryos and fetuses encased in sacs, which he mistakenly identified as eggs. The egg, Harvey felt, contained the makings of "all that is alive." Semen was relegated to the role of a "contagion," prompting human generation much as a virus does a cold.

And how would the life force, the soul, get into the egg? Here science abandoned Harvey, and he fell back on religion: "It is given . . . by the heavens or the sun or the Almighty Creator."

Like most biologists of his day, Harvey was limited by his equipment. He couldn't see what was going on at the cellular level. He couldn't see sperm. He had a magnifying glass, but what he needed was a microscope. And so it is not surprising that the next milestone fell to Antoni van Leeuwenhoek: the man who loved microscopes. Leeuwenhoek did not invent the microscope, nor was he a scientist by training. The Dutchman worked as an accountant for a haberdasher and, later, as Chamberlain of the Council-Chamber of the Worshipful Sheriffs of Delft, which is a ten-gallon way of saying that he tidied the chamber. The post left him plenty of time for hobbies, of which he had just one: grinding lenses and building microscopes. The microscopes Leeuwenhoek built were superior to those of the main maker of the day, and soon they were in demand by members of the Royal Society of London for the Improving of Natural Knowledge, more or less the National Science Foundation of its day. Over time, the Royal Society began publishing Leeuwenhoek's letters about his findings as well, and he was on his way to a historical, if unpaid, career as the founding father of microbiology.

In 1675, Leeuwenhoek discovered a universe of up-to-then-unknown creatures—bacteria and protozoa, mostly—in drops of stagnant water in a "water-butt" in his yard. He named them animalcules. It is difficult to properly appreciate the wonder and strangeness of this discovery. Think of scientists today discovering Martian life. Leeuwenhoek was appropriately awed. "For me this was among all the marvels that I discovered in nature the most marvelous of all, and I must say, that for my part, no more pleasant sight has met my eye than this of so many thousands of living creatures in one small drop of water."

Leeuwenhoek bravely turned his instrument upon himself. "My teeth are not so cleaned . . ." he wrote, "but what there sticketh or groweth between some of my front ones and my grinders . . . a little white matter, which is as thick as if 'twere batter." He mixed some of this batter with fresh rain-water and looked at a smear under the microscope. Did he find animalcules? You bet your water-butt he did. "All the people living in our United Netherlands," he concluded, "are not as many as the living animals that I carry in my own mouth." It is a testament to Leeuwenhoek's love of biology that he could describe the bacteria in tooth scum as "very prettily a'moving."

In a further exploration of oral fauna and the limits of spousal patience, Leeuwenhoek headed into the mouth of his wife, Cornelia, and their daughter Maria. "I examined . . . a little of the matter that I picked out with a needle from betwixt their teeth." Next he recruited an old man who had "never washed his mouth in all his life" and noted that while his spittle held a normal number of these animalcules, the matter between his teeth held "an unbelievably great company of living animalcules." Day by day, the foundations of modern oral hygiene took shape under Leeuwenhoek's lens. He noted the relationship between a "stinking mouth" and "the animals living in the scum on the teeth." In a three-hundred-years-premature dig at Listerine, he observed that while wine-vinegar killed spittle animalcules on contact, it "didn't penetrate through all the matter that is firmly lodged between the front teeth or the grinders and killed only those animalcules that were in the outermost parts of the white matter."

While the fellows of the Royal Society were politely attentive to Leeuwenhoek's oral safaris, they encouraged him to move on to the rest of man's moistnesses. In particular, they wanted him to examine semen. Perhaps it would be possible at last to view the material of the human soul! Leeuwenhoek refused. "He questioned the propriety of writing about semen and intercourse," wrote E. G. Ruestow in an article in the

Journal of the History of Biology. Several years later, a medical student presented Leeuwenhoek with a vial of semen from a gonorrheaic man. (Hey, *thanks*!) The student said he'd found within it small animals with tails, which he assumed were related to the gonorrhea. Leeuwenhoek suspected otherwise, and set about examining his own semen. In a 1677 letter describing his findings, Leeuwenhoek was careful to point out that the material was a "residue after conjugal coitus," and not the product of "sinfully defiling myself."

In that letter Leeuwenhoek sets forth the first scientific description of sperm: animalcules so small that "a million of them would not equal in size a large grain of sand. . . ." He describes the apparent difficulties of swimming in semen, noting that the animals had to "lash their tails eight or ten times before they could advance a hair's-breadth." He included eight drawings of "the little animals in the seed," some with tails straight, looking like hat pins, others with winding sine-wave tails, clearly struggling against the custardy tide.

Then he commenced to tread the path that would lead to his biggest career blunder. He claimed to see a network of vessels within the sperm bodies, and imagined that it held the beginnings of all the organs that the human would one day possess. This line of thinking—called preformationism—would prove enormously popular and to this day provides publishers of embryology textbooks with irresistible images for their historical chapters: old woodcuts and engravings of sperm with microscopic humans inside, heads down and knees drawn up to their chests, like cramped, napping stowaways. One of these likely influenced Leeuwenhoek. He had received a letter from a French aristocrat named François de Plantade, which included two drawings of the miniature people inside sperm. In this case, they were depicted outside of their sperm hulls, standing with their hands crossed demurely over their little private parts. On their heads are what appear to be small hats or hooks, giving them the appearance of adorable human bracelet charms.

Though Leeuwenhoek himself never managed to find the preformed people inside sperm—despite having tried at one point to peel the "skin" off one—he came to believe they were in there. He believed that each sperm held a soul with the potential to become a human life, and that the woman's role in reproduction was merely to receive and nourish the perfectly formed miniature human. (Leeuwenhoek wasn't the first to espouse this line of thought. Hippocrates took the no doubt breakfast-inspired view that the egg was simply something for the developing human to eat. He further speculated that as soon as the egg was all eaten up, then the infant would hatch: birth as a sort of grocery-shopping trip.)

Leeuwenhoek was what became known as a spermist. The label suggests that there were ovists for the spermists to argue with over dinner, and indeed there were. I learned about the ovist-spermist debate in an amazing book called *The Ovary of Eve*, by Clara Pinto-Correia, who has the audacity to be both a literary success and a respected developmental biologist. I don't know what Clara hatched from, but clearly better stuff than I.

The ovists pointed to the spherical shape of the ovum as befitting its lofty mission. The sphere is the shape of the planets and the stars: God's perfect form. (Whereas sperm look like worms.) Leeuwenhoek took a different view. He didn't think of ova as spheres; he thought of them as globules. "Do we not see that all excrements, discharged either by human beings or animals, consist of globules . . . ?" he wrote. "And . . . we see that fat, pus, and certain parts of a horse's urine also consist of globules." This from the man who didn't want to write about semen.

The ovum's main shortcoming as the vessel of humanity was that it derived from the woman, who was in those days (more) universally considered a second-class organism. "If ovism was the true system of reproduction, God was sending a mixed message," writes Pinto-Correia. "He had locked us inside perfection. And then he had locked perfection inside imperfection."

The other argument for the primacy of sperm was that they moved. They appeared to possess some kind of animating spirit. On the other hand, if sperm was an animal, did that mean that it ate and defecated and copulated? Pinto-Correia's book includes a detailed drawing by an overimaginative French embryologist purporting to show the wee digestive system of the human sperm. But if sperm hatched humans, who or what hatched sperm? Not surprisingly, there were competing theories as to sperm's function. Some thought sperm had nothing to do with reproduction, and guessed them to be a symptom of testicle disease. Others thought the wrigglers' job was to incite the male into having sex, presumably by causing some kind of physical itch or discomfort, though I like to envision a sort of Woody-Allen–style team effort involving tiny megaphones and shouts of encouragement.

The debate dragged on until Oskar Hertwig came along to set things straight. Following the discovery of conception, the question of the hour became: When does the soul enter this new being, this cellular amalgam of male and female? Conception—the mystical fusing of egg and sperm—was the logical choice, quickly supplanting Aristotle's notion of the evolving soul.

Contemporary debate over the morality of abortion and stem cell research has sparked renewed interest in the timing of human ensoulment. The best book I found on the topic is a Cambridge University Press publication by Norman Ford, called *When Did I Begin?* Ford, a moral philosopher and a Salesian Catholic priest, makes the clean and quite elegant argument that personhood—to use the more secular term for ensoulment—cannot begin until after the point where identical twinning is no longer possible: about fourteen days after conception. Up until that point, it's possible for the zygote to become two identical twins. If the soul had arrived at conception, what would happen then? Would it split into two, each twin making do with half a soul? No, Ford argues. Up until that point, the zygote—with its potential to become two distinct and separate human beings—cannot rationally be referred to as a person. "I contend that the cell cluster can best be understood as human biological material but not a unified living human organism," he writes.

As for exactly what point after the fourteenth day personhood might begin, that is less clear. The fourteenth or fifteenth day heralds the arrival of the primitive streak, the early vestiges of the neurological system, and some argue for this point. But no one—at least on a scientific basis—knows for sure when the soul, the spirit, the self, is instilled, or installed, or whatever process it uses to get itself in there. Or what it consists of or where it's located. Or even if it exists. Which brings us back to our basic quest.

> Descartes became a familiar sight at the butcher shops in Amsterdam, where he would buy freshly slaughtered animals. When visitors asked to see his library, he would take them into a room where he kept carcasses in various stages of dissection. "These are my books," he would say.

The above passage, from science writer Carl Zimmer's *Soul Made Flesh*,* describes one of philosopher René Descartes's lesser-known projects: to figure out the workings of the human machine. One of the specific things Descartes was doing with his carcasses was looking for the soul. He assumed it resided somewhere in the brain, and so his most well-thumbed "books" took the form of cow heads. One paragraph up, Zimmer writes that Descartes spent much of that period of his life in self-imposed exile, "craving solitude." The carcasses surely helped.

Descartes is one of the few early philosopher/scientists to have physically searched for the soul, actually opened up bodies and looked for it. He eventually nominated the pea-sized pineal gland. To those who know the gland's actual function (it regulates melatonin production), it may seem an unlikely choice. Descartes was swayed by the gland's position at the center of the head, and by dint of its being one of the few brain structures that don't exist in pairs. He didn't think the ugly little gland *was* the soul per se; more that it was a sort of hub, a meeting point for sensory information and the flowing streams of spiritus (akin to Aristotle's pneuma) that carried out the self's higher functions.

Descartes dreamed up an elaborate model of the nervous system with strings and valves and tiny bellows. He described spiritus flowing through the nerves—which he envisioned as tubular—and into the muscles, causing them to contract by inflating parts of them. In a paper called "On the 'Seat of the

---

*Zimmer's book is about the dawn of neuroscience: the first men to open up heads and figure out how brains worked. Zimmer once edited a story of mine for *Discover*, a situation from which he's probably still recuperating. The guy is smarter than anyone I know. If you were to open up his head, his brain would burst out like an airbag.

Soul': Cerebral Localization Theories in Mediaeval Times and Later," neurologist O. J. Grüsser writes that Descartes's model for this system was the organ, then in its heyday as a popular musical instrument. Fifteen centuries earlier, Grüsser adds, the Greek physician Galen based his system of spiritus flow on the mechanics of Roman bathhouse heating systems. Meanwhile, the philosopher Albertus Magnus found inspiration in the equipment used to distill brandy. And so it went, on into the twentieth century, when tape recorders and computers took hold as the working models of consciousness.

A couple years back, I corresponded with a computer professional named Betty Pincus, who has been in the industry some forty years. "It always interested me the way some of my colleagues would use the technical vocabulary to describe how their minds worked," she wrote. "In the sixties, they talked about 'running out of tape' or 'her accumulator overflowed.' As the technology changed, it became 'running out of disk space' or 'multitasking.' I've often wondered whether the inventors of these machines created them in their own image of how their minds worked or if they related the machine to the mind after the machines were created."

I managed to find only two other references to scientists rummaging around in corpses looking for souls. One comes from the Midrash, a collection of ancient rabbinical commentaries on the Torah. The Midrash makes reference to a single indestructible bone, called the luz. The luz is shaped like a chickpea (or an almond, depending on which rabbi you listen to) and located at the top of the spine (or the bottom, depending on which rabbi you listen to). From this bone, the Midrash states, a person is reconcocted after death. It's the soul bone.

The Midrash includes a description from the Torah of an experiment to prove the unique indestructibility of the luz.

The project was carried out by Rabbi Joshua ben Hananiah upon being confronted by the Roman emperor Hadrian. "Prove it to me," Hadrian is quoted as saying. So the rabbi did. "He had one brought. . . ." (Like he just turned to some underling: *Smetak! Fetch me a luz!*) "He put it in fire, but it was not burnt, he put it in a mill but it was not ground. He placed it on an anvil and struck it with a hammer; the anvil split and the hammer was broken but all this had no effect on the luz."

Needless to say, there was much chatter about the luz over the centuries, and when the science of anatomy began to gain momentum in the Middle Ages, its practitioners sought to find it. They nominated, among other bones, the coccyx, the sacrum, the twelfth dorsal vertebra, the wormian bones of the skull, and the tiny sesamoid bones of the big toe. Of course, these bones are all easily destroyed, and the anatomists eventually decided it was a matter best left to the philosophers. The famed Renaissance anatomist Vesalius, having spent an afternoon mucking around with a set of sesamoid bones, more or less laid the matter to rest: "We should attach no importance whatever," he wrote in *De humani corporis fabrica*, "to the miraculous and occult powers ascribed to the internal ossicle of the right great toe."

Contemporary rabbinical discussion of the luz is harder to come by. On the website Ask the Rabbi, a mohel from Paris posted an e-mail seeking information about the luz. The rabbi's reply confirmed the bone's alleged indestructibility and added that it has been described as "having within it many intertwined spider-like blood vessels." He referred the mohel to a Dr. Eli Temstet of Paris for more information. I e-mailed the mohel to see what he'd found out. "Dr. Eli Temstet of Paris has gone to a better world," came the reply. "Now, he sure knows where and what exactly is the luz." I sent an e-mail to Ask the Rabbi. Was there a paper on the spider-like blood vessels of the luz? The rabbi did not Answer the Writer. I consulted a book on Talmudic medicine, but no mention was made of the spidery veins of the soul bone.*

The very first person to poke around for a soul in a human cadaver was the third-century B.C. physician Herophilus, of Alexandria. Herophilus is thought to be the first person in history to have dissected human cadavers for the purpose of scientific enlightenment. As such, he bagged a lot of anatomical discoveries. One of these was the four chambers, or ventricles, of the brain. He believed that the soul was headquartered in the fourth one. Why was Herophilus looking around for a soul inside a dead man, especially given that Egyptians believed in an afterworld† for souls to retire to? I can't tell you for sure, but I can tell you that Herophilus was rumored to have dissected live humans as well. Two colleagues accused him of vivisecting hundreds of criminals. Perhaps his soul-related aspirations explain his poor table manners.

If your goal was to pinpoint the soul, it obviously made

---

*My disappointment was short-lived, for this was a wondrous book. Here were detailed rabbinical opinions upon "whether or not a cattle breeder whose animal caused damage by knocking something with its penis must make restitution" (undecided); upon the inadmissibility of cleansing the anus "with the snout of a dog"; upon "the misconduct in which a woman places into the vagina of another woman a piece of meat from a fallen animal." Here were descriptions of "hairy heart" and treatments for chronic uterine bleeding ("take three measures of Persian onions, boil them in wine, make her drink it and say to her, 'Cease your discharge!' ").

†A rather barren place, from what I gather. Egyptians made frequent trips to the family plot to supply departed souls with food, clothing, and kitchen items. According to Clara Pinto-Correia, some tombs were even outfitted with toilet facilities for the ka (soul). That No. 2 carries over into the afterlife was apparently a common belief. Correia cites a reference to a funerary fragment expressing anxiety over the possibility that the ka, should its food cache run out, might resort to feeding on its excrement.

more sense to experiment on the living than on the dead. The simplest plan of action would be to systematically scramble, excise, or otherwise disable the likely structures and watch to see if the lights went out. And this, more or less, is what got done. Like Descartes, most scientists had zeroed in on the brain. (From early on, observations of personality changes caused by head injuries suggested a link between brain and self.) Galen was one of the earliest of the neuro-vivisectors; he experimented with cutting (and getting on) the nerves of his neighbors' pigs. Based on this work, he decided that the soul was situated in the substance of the brain and not, as Herophilus had maintained, in the ventricles.

Leonardo da Vinci further narrowed it down. In 1996, McGill University professor of neurosurgery Rolando Del Maestro curated an exhibition called "Leonardo da Vinci: The Search for the Soul." In the materials for the exhibit, Del Maestro describes a 1487 manuscript in which Leonardo made notes about a passage he had read in a book about the Carthaginian War. The passage describes the quickest way to kill an injured elephant: by pounding a stake between the animal's ears at the top of the spinal column. Intrigued, Leonardo took to pithing frogs in a similar manner. "The frog instantly dies when its spinal medulla is perforated," Del Maestro quotes Leonardo as having written. "And previously it lived without head, without heart or any interior organs, or intestines or skin. Here, therefore, it appears, lies the foundation of movement and life." (I think that what he meant is that the skinned, gutted, headless, heartless frogs lived *for a little while*.)

Only one soul-seeking man of science carried out this sort of cavalier slice-and-see experimentation on a living human being. As the king's surgeon and the founder of France's Royal Academy of Surgery, Gigot de La Peyronie could pretty well do what he felt like. In 1741, he published a paper entitled

"Observations by Which One Tries to Discover the Part of the Brain Where the Soul Exercises Its Functions." The subject was a sixteen-year-old boy whose skull had been cracked by a rock. After three days of worsening symptoms, the youth fell unconscious. La Peyronie "opened up his head" and found a suppurating abscess deep down inside the brain, at the corpus callosum. He drained the wound, taking care to measure the runoff, which amounted to "about the volume of a hen's egg." As soon as the pus* that had weighed upon the corpus callosum was drained, he wrote, the coma lifted. La Peyronie noted that when the cavity had refilled with ooze, the youth fell unconscious again. He reemptied it, and again the boy awoke. The corpus callosum, he reasoned, must be the seat of the soul. Just to be sure, La Peyronie decided to undertake a little experiment. He filled a syringe with saline and injected it directly into the newly drained wound. As predicted, the boy lost consciousness. And was brought back to his senses when La Peyronie pumped the water back out.

La Peyronie found confirmation of his conclusions in three patients who had lost consciousness and died following head injuries and then been found during autopsy to have abscesses in the vicinity of the corpus callosum. One of the abscess pockets, that of a soldier whose horse had kicked him, was here again described as being "of the size and shape of a hen's egg"—clearly the "bigger-than-a-breadbox" of its day.

Autopsies also enabled La Peyronie to rule out Descartes's competing claim that the soul hung its hat in the pineal gland. La Peyronie had autopsied patients discovered to be missing

---

*I was intrigued to learn that the French for "pus"—even among members of eighteenth-century aristocracy—is "*le pus.*"

the gland entirely or in whom it appeared to have petrified. "*Elle ne réside pas dans la glande pinéale*," he declared. I find a certain arrogance suffuses both La Peyronie's writing and his deeds. If only he could know that today, as regards historic Frenchmen known to Americans, La Peyronie doesn't hold a flame to Le Petomane,* the Moulin Rouge "fartiste," whom no one should hold a flame to anyway.

La Peyronie was the last of a breed—a lone holdout in the anatomical search for the soul. Most of the early neuroanatomists had come to see that the self was too complex, too multifaceted, to be housed in or operated by a single biological entity. Like the Soviet Union after Gorbachev, the once broad and uniform-seeming soul began to splinter into dozens of smaller republics. Through a combination of unpleasant and often contradictory animal studies—living brains pokered, ablated, and hacked—and autopsy studies that sought to match brain abnormalities with dominant personality features, the men of science began mapping the specialized duties of the brain's real estate, a project that continues to this day.

Of all the brain's early cartographers, none was quite so thorough as Viennese physician Franz Joseph Gall. Gall claimed to have located twenty-seven distinct "organs" of the human brain, each corresponding to a specific trait or faculty. By all accounts a gifted physician and anatomist, Gall succeeded in pinpointing the brain's language center and that of our memory for words.

---

*I feel it would be wrong to introduce Le Petomane into a manuscript and then abandon him in the very same sentence. I had always thought that the act consisted of popular songs performed on his own wind instrument. But I learned from "The Straight Dope" columnist Cecil Adams that, in fact, Le Petomane, whose real name was Joseph Pujol, could produce only four notes without the aid of an ocarina. This is not to belittle his rectal talents. Pujol could smoke a cigarette down to its butt (or his butt, or both) and blow out candles, as well as expel a fountain of water several feet into the air.

His other "organs" were rather more questionable. For instance: The Organ of Poetical Talent. The Organ of Metaphysics. And, my personal favorite: The Organ for the Instinct for Property-Owning and Stocking Up on Food. Gall's organs landed him in hot water with the church, which labeled him a heretic for teaching that man had multiple souls, a charge Gall denied.

Gall was led astray in part by his unconventional methodology. The swift decomposition of brains precluded their lengthy study, so Gall took to examining skulls, both of the living and the dead. He reasoned that if an organ of the brain was particularly well developed, it would put pressure upon the cranium and raise a bump that could be seen or felt through the hair. (Phrenology—a mass-market popularization of his theories—had the masses feeling each other's heads and galling Gall for decades to follow.) Gall amassed a collection of 221 skulls, which traveled with him on his lecture circuit, exasperating porters and alarming nosy bellhops. He also owned, at last count, 102 plaster casts of human heads, many of which he'd made himself. The heads were casts of people he met in his travels whose character seemed obviously dominated by one or two strong traits and whose skull bore a bulge in the appropriate spot: evidence for his theories. Skull #5491, for instance, belonged to a Mr. Weilamann, the director of a portable hydrogen gas generator* company, and showed a notable bump over the Organ for Mechanical Sense, Construction, and Architecture.

---

*In looking up "portable hydrogen gas generator" on Google, I came across a study called "Detection of Flatus Using a Portable Hydrogen Gas Analyzer," apparently a novel use of the device. The author taped the machine's sampling tube to twenty postoperative gastrointestinal patients' buttocks in an effort to detect farts, a happy sign that their plumbing was back in action. Hydrogen is the main component of flatus; you and I are, in essence, hydrogen gas generators of a less portable variety.

Gall was quite devoted to his collection. To track down examples of the Organ of the Penchant for Murder and Carnivorousness, for instance, he took to wandering through prisons, looking for murderers with ridges above the ears. Lunatic asylums were another fruitful stop for Gall and his plaster craft. The catalogue of Gall's collection contains dozens of items like #5494: "Copy in plaster of the skull of a total idiot."*

For the Organ of Poetical Talent, Gall resorted to fondling marble busts of the great poets. As evidence for the location of the Organ of Belief in the Existence of God, he cites a series of Raphael paintings in which Christ appears to have a noticeable rise at the crest of his cranium, as though Satan had bopped him over the head with his trident. Had Gall gone potty? Possibly. Here is his evidence for the Organ of the Instinct of Propagation. He knew of a young clockmaker who, when he "ejaculated by onanism," would lose consciousness for an instant and suffer convulsive movements of the head and a violent pain in the back of the neck. "The idea couldn't escape me," writes Gall in *Sur les fonctions du cerveau*, "that there must be a connection between the functions of physical love and the cerebral parts in the nape of the neck." Then again, perhaps there is a connection between violent convulsive head movements and neck pain.

As further evidence for the Organ of the Instinct of Propagation, Gall cites a young widow who admitted that

---

*The terms "idiot" and "lunatic" were acceptable diagnostic terms in England up until 1959. "Imbecile" and "feeble-minded person" were, likewise, listed as official categories in the 1913 Mental Deficiency Act. England has always lagged a bit behind in discarding outdated terms for the disadvantaged. (When I was there in 1980, it was still possible to shop for used clothing at the local Spastic Shop.) That is, compared to the United States, where it takes, oh, about twenty-five minutes for a diagnostic euphemism to become a conversational faux pas.

since childhood she had felt "strong desires that were impossible to resist" and during these moments the nape of her neck burned. Gall describes placing his hand on her young widowly nape during one of these burning-desire episodes and discovering "a very considerable rounded prominence," possibly one of several going on in the room.

Item #19.216 of the Gall collection is the skull of Franz Joseph Gall. Gall disciple N. J. Ottin notes that "on the occiput, the tendency toward sex was very marked."

From Gall's day onward, the soul began to drift away from the provinces of anatomy and neurology and off into airier domains: religion, philosophy, parapsychology. The men of medicine were through with the soul—with one terrifically odd exception.

3

How to Weigh a Soul

What happens when a man (or a mouse, or a leech) dies on a scale

IT WAS A pretty place to die. The mansion on Blue Hill Avenue was the showpiece of the Dorchester, Massachusetts, estate known as Grove Hall. Four stories tall, with a porticoed porch and cliques of indolent shade trees, the mansion had been home to T. K. Jones, a wealthy merchant in the China trade. In 1864, it was bought by a physician-cum-faith-healer named Charles Cullis, who turned it into the Consumptives' Home—a charitable operation for late-stage tuberculosis (a.k.a. consumption) patients. With the discovery of antibiotics sixty years off, prayer was as useful a treatment as any then on offer. TB patients were routinely packed off to sanitariums, ostensibly to partake of rest "cures," but mainly to keep them from spreading the disease.

Had you been visiting the Consumptives' Home in April 1901, you might have been witness to a curious undertaking. A plump, meek-looking man of thirty-four, wearing wire-frame glasses and not as much hair as he once did, was stooped over the platform of an ornate Fairbanks scale, customizing the device with wooden supports and what appeared to be an army-style cot. The scale was an oversized commercial model, for weighing silk—no doubt a holdover from Jones's mercantile days.

Clearly something unorthodox was afoot. Though weight loss was a universal undertaking at the Consumptives' Home, no one needed a commercial scale to track it.

The man with the hammer was Duncan Macdougall, a respected surgeon and physician who lived in a mansion of his own, in nearby Haverhill. Macdougall was acquainted with the Consumptives' Home attending physician, but he himself was not on staff. Nor was he treating any of the patients, or even praying for them. Quite the opposite; Macdougall was literally—perhaps even a little eagerly—waiting for them to die.

For the preceding four years of his life, Duncan Macdougall had been hatching a plan to prove the existence of the human soul. If, as most religions held, people leave their bodies behind at death and persist in the form of a soul, then mustn't this soul occupy space? "It is unthinkable," wrote Macdougall, "that personality and consciousness can be attributes of that which does not occupy space." And if they occupy space, he reasoned, they must have weight. "The question arose in my mind: Why not weigh a man at the very moment of death?" If the beam moved, and the body lost even a fraction of an ounce, he theorized, that loss might represent the soul's departure.

Macdougall enlisted the help of two fellow physicians,

Drs. Sproull and Grant, who chose not—or possibly weren't invited—to put their names on the research paper. The plan was to install a cot on the scale platform and then install a dying consumptive on the cot. Death from consumption is a still, quiet affair, and so it fit Macdougall's conditions "to a nicety," as he put it. "A consumptive dying after a long illness wasting his energies, dies with scarcely a movement to disturb the beam, their bodies are also very light, and we can be forewarned for hours that a consumptive is dying." I found his enthusiasm at once endearing and a little troubling. I imagined him addressing the ward as he canvassed for volunteers. (Macdougall wrote in the *Journal of the American Society for Psychical Research* that he secured his subjects' consent some weeks before their deaths.) *You people are just perfect for this project. A, You're easy to lift, B, you're practically comatose when you go. . . .* Who knows what the consumptives made of it, or whether they were too out of it to know what he was asking.

At 5:30 p.m. on April 10, 1901, Patient 1's death—"my opportunity," Macdougall called it—was declared imminent. A male of ordinary build and "standard American temperament," he was wheeled from the ward and lifted onto the scale like a depleted bolt of silk. Macdougall summoned his partners. For three hours and forty minutes, the physicians watched the man fade. In place of the more usual bedside attitudes of grief and pity, the men assumed an air of breathless, intent expectancy. I imagine you see this on the faces of NASA engineers during countdown and, possibly, vultures.

One doctor watched the man's chest; another, the movements of his face. Macdougall himself kept his eyes on the scale's indicator. "Suddenly, coincident with death," wrote Macdougall, "the beam end dropped with an audible stroke hitting against the lower limiting bar and remaining there with no rebound. The loss was ascertained to be three-fourths of an ounce." Which is, yes, twenty-one grams. Hollywood metricized its reference to the event for the simple reason that *21 Grams* sounds better. Who's going to go see a movie called *Point Seven Five Ounces*?

Over the years, Macdougall repeated the experiment on five more patients. A paper summarizing his findings ran in the journal *American Medicine* in 1907. In the months that followed, dubious M.D.s launched their criticisms in lengthy letters to the editor. Macdougall countered them all. One correspondent pointed out that the sphincter and pelvic floor muscles relax at death, and that the loss was perhaps urine and/or feces. Macdougall patiently replied that if this were the case, the weight would remain upon the bed and, therefore, upon the scale. Someone else suggested that the dying patients' final exhalation might have contributed to the drop in weight. To prove that it hadn't, Macdougall gamely climbed onto the cot and exhaled "as forcibly as possible," while Sproull watched the scale. No change was observed.

The most likely culprit was something called "insensible loss": body weight that is continually being lost through evaporating perspiration and water vapor in one's breath. Macdougall claimed to have accounted for this. His first patient, he wrote, lost water weight at the rate of an ounce per hour, far too slowly for insensible loss to explain the sudden three-quarter-ounce drop at death.

THE HISTORICAL AUTHORITY on insensible weight loss is a Paduan physiologist named Sanctorius. Known humdrumly as the "founding father of metabolic balance studies," Sanctorius coined the term "insensible perspiration" in 1690, in a diverting volume entitled *Medicina statica.** To aid him in his research, Sanctorius devised an experimental scale of his own. He suspended a platform on a massive steelyard scale. The platform held a bench with a hole cut out of the center of it and a bucket underneath it, and in front of the platform stood a supper table: Out box and In box. Sanctorius sat himself down on the platform, enjoyed a meal, and then sat around on the scale for eight hours, availing himself of the bucket when needed. He then weighed, to use his exuberantly capitalized phrasings, "the Excrements of the Guts"—observing on an unrelated tangent that "the thick ones are lighter and *swim.*" Sanctorius found that a small portion of the food weight remained unaccounted for, i.e., wasn't down there in the bucket. This he ascribed to evaporated sweat and breath vapor, which he collectively dubbed insensible perspiration.

Sanctorius calculated that an eight-pound intake of meat and drink will, over one day, yield five pounds of insensible perspiration—or an average of three ounces of sweat and breath vapor lost per hour: three times the rate Macdougall observed. At one point Sanctorius describes the digestion of "a supper of eight pounds."† It soon became clear there was little overlap between the dripping trencherman of Sanctorius's day and Macdougall's dry little consumptives. I skipped ahead to Section VI, which was all about the effects of immoderate coitus on insensible perspiration. Sanctorius effected the

---

*Medical treatises were eminently more readable in Sanctorius's day. *Medicina statica* delves fearlessly into subjects of unprecedented medical eccentricity: "Cucumbers, how prejudicial," "Phlebotomy, why best in Autumn," and the tantalizing "Leaping, its consequences." There's even a full-page, near-infomercial-quality plug for something called the Flesh-Brush.

†Astoundingly, Sanctorius was described as a small man. His work habits may explain his ability to stay slim in an era of eight-pound dinners. He claimed to have tested ten thousand subjects over twenty-five years.

quaint habit of presenting his findings in the form of aphorisms. As in, "Aphorism XXXIX: Such a Motion of a Body as resembles that of a Dog in Coition, is more hurtful than a bare Emission of Semen; for the latter wearies only the internal Parts, but the other tires both the Bowels and the Nerves." Or "Aphorism XL: To use Coition standing, after a Meal, is hurtful; because as it is upon a full Meal, it hinders the Offices of the Bowels." Sanctorius preached that by obstructing insensible perspiration, immoderate sex led to everything from "Palpitations in the Eyebrows and Joynts" to a hardening of the tunicles of the eyes—and here we have what I surmise to be the original striking of the masturbation-makes-you-go-blind myth. Sanctorius preached a carnal moderation that seemed almost killjoy—all the more so for the book's wanton promotion of oysters as sources "of the greatest possible nourishment."

TO GET TO THE bottom of the insensible weight loss conundrum—is it an ounce per hour, as Macdougall calculated, or is it three?—I called America's modern-day Sanctorius, Eric Ravussin. Ravussin, currently with the Pennington Biomedical Research Center in Baton Rouge, used to run metabolic chamber studies for the National Institutes of Health. He, too, has measured insensible water loss during sleep—by tucking volunteers into beds on platform scales inside the chamber. His findings came in right beside Macdougall's: about an ounce per hour. Macdougall was right: It would be hard to imagine insensible water loss as the force behind an instantaneous drop of three-fourths of an ounce.

Ravussin had no idea what could have caused the abrupt weight loss. He referred me to a book by Max Kleiber, called *The Fire of Life: An Introduction to Animal Energetics.* Though a tad formula-heavy for the likes of me, the book is as diverting in spots as was Sanctorius's. We learn, for instance, that the "extra-large vagina of the Brahman cow is an effective organ for heat dissipation." In a similar vein, Schmidt-Nielsen "observed that a camel's rectal temperature may rise during a day from 34.2 to 40.7 C," though I doubt that, strictly speaking, observation alone did the trick. Sometimes you have to get right in there, as Kleiber himself did in 1945, calculating "the insensible weight loss of cows in pasture by preventing their water and food intake with a muzzle and collecting and weighing all feces and urine." I skimmed the entire book, looking for some reference to a sudden weight drop at death. I found nothing. There is only so much one can do. In the words of Max Kleiber, "If we insisted on meeting all our fuel needs with eggs, we would soon reach the end." Or something.

SO HOW ARE we to explain Macdougall's befuddling finding? I have some theories for your consideration.

Theory the First: Duncan Macdougall was a nutter. I was an early supporter of the nutter theory, based largely on the fact that Macdougall was a member of the Massachusetts Homeopathic Medical Society. He wrote his medical school thesis on the Law of Similars, the underlying principle of homeopathy—basically Like Cures Like. I don't know what homeopathists get up to nowadays, but back in the movement's infancy it was nutter central. The homeopathists' bible, *A Dictionary of Practical Materia Medica*, is a three-volume compendium of plants, animals, and minerals, and the symptoms they produce if you ingest them, which homeopathists did a lot of, perhaps accounting for the nutter situation. The central tenet was that substances that cause healthy people to get certain symptoms can cure diseases with these same symptoms. The early homeopathists spent years dosing themselves and their patients and friends with every substance they could get hold of, and carefully cataloguing the reported symptoms. I can't vouch for the movement's contributions to the healing arts—without control groups or placebos, the *Materia* work is meaningless by modern research standards—but I must commend their flair for language. For example, we have alumina causing "dreams of horses, of quarrels, of vexations" and a "tingling on the face, as if it were covered with a white of egg dried." Agnus castus causes "odor before the nose, like herrings or musk," as well as "feeble erections" and—it almost goes without saying—"great sadness." And then there is chamomile, said to cause the symptom "cannot be civil to the doctor."

But at the time Duncan Macdougall went to medical school, in 1893, homeopathy was not considered a fringe branch of medicine. About half the country's medical schools—including Macdougall's alma mater, Boston University—still taught the homeopathic approach to healing. (BU had dropped it by the early 1920s.) The point is, plenty of mainstream, straight-ahead physicians practiced homeopathy in Macdougall's day.

Also working against the touchy-feely flake theory are the plentiful examples of Macdougall's consistent toe-the-line geekdom. He was class president and class orator at BU. A 1907 article in the *Boston Sunday Post* flatly stated that Macdougall was a believer in neither spiritualism nor psychic phenomena. A *Haverhill Evening Gazette* piece described him as "hard-headed and practical." Greg Laing, head of the History Room at the Haverhill Public Library, recalls visiting the Macdougall household with his parents as a boy, so I asked him about the good doctor's nutter potential. (Macdougall had died by then, but his widow and son were still living.) "God, no," said Laing. "They were such grim, straitlaced people. Really and truly, they were not esoterically inclined." I phoned Olive Macdougall, the widow of Macdougall's only grandchild. Though her husband never knew his grandfather, Olive confirmed the family's decidedly nonmystical bent. Her father-in-law, Duncan's son, was a banker and lawyer.

The writer of Duncan Macdougall's *Gazette* obituary tried to foist a little jollity on the man, but it was a thin effort: "He was cheerful in the sickroom and some of his sickroom phrases and words of encouragement remain on the tongues of his patients. A few of his sickroom phrases were: 'Don't you worry, my gal, everything will be all right' and 'Don't you worry and you'll get well in a bigger hurry.'"

Macdougall was neither madman nor visionary. What he was, I'm guessing, was a henpecked little man in need of attention. Greg Laing described Macdougall's wife Mary as "a battle-axe of monumental proportion." (Perhaps a chamomile tea drinker.) "I don't think she had the slightest respect or interest in her husband's project." Macdougall got his strokes from his work. As far as I can tell, he made a habit of calling up the local papers to garner laurels where he could. "Dr. Macdougall Becomes Poet," overstates the headline when some limp doggerel ran in *Life*. "Dr. Macdougall Wins Great Fame," blusters another, after England's navy agreed to have its Royal Marine Bands play Macdougall's lurching composition "The British Tar's Song." (Macdougall's nephew had a contact at the Admiralty, whom he deluged with 1,800 copies of the song.)

Theory the Second: Macdougall's experimental protocols were as lame as his poetry. Let's look a little closer at his findings.

Macdougall weighed six patients in all, but only the first, the one described earlier, stands as a strong example of the phenomenon. Macdougall threw out Patient 6's data because the man died just as they had put him on the cot and were adjusting the beam. Number 4's data he discounted because, Macdougall wrote in *American Medicine*, "our scales were not finely adjusted and there was a good deal of interference by people opposed to our work." The doctor makes several references to "friction on the part of officials," and states that only the first patient was run under ideal conditions, i.e., sans friction. He doesn't specify what form this friction took, but if it affected the tests to the extent that some were thrown out, it seems reasonable to assume that the officials were there in the room, hectoring Macdougall or trying to bring a halt to what he was doing. Hardly ideal conditions for a test that requires concentration and enough quiet to listen for a heartbeat.

That leaves four subjects. With the exception of Number 1, the data for all were compromised in one way or another. Number 2 stopped breathing at 4:10 a.m., but the scale didn't budge for another fifteen minutes (whereupon it registered a half-ounce drop). "We had great doubt, from the ordinary evidence," writes Macdougall in his *American Medicine* paper, "to say just what moment he died." If you can't tell when the man died, you can't very well claim that he lost a half ounce at the moment of death.

Number 3's weight loss happened in two phases: a half-ounce loss at the moment of death, and then an additional loss of an entire ounce a few minutes later. Macdougall explains that the second loss might have been caused by a jarring of the scale, caused when one of his colleagues listened to the subject's heart. If pressing a stethoscope to a patient's chest disturbs the balance, as of course it would, then how did Macdougall and his colleagues presume to know the moment of death in *any* of these cases?

Number 5's data were tainted by a peculiarity of the scale.

Following a three-eighths-ounce loss, three-eighths ounces of weight was added to the scale to bring it back to zero; however, the beam didn't budge for fifteen minutes. Macdougall had no explanation. Was his scale dodgy? Did Fairbanks make a reliable scale? Was it really accurate to one-fifth of an ounce? Where was a Fairbanks scale historian when you needed one?

PEGGY PEARL OVERSEES the Historical Collection of the Fairbanks Museum in St. Johnsbury, Vermont, where Franklin Fairbanks began manufacturing scales in 1830. The collection includes 30 or so antique Fairbanks scales, as well as farm implements, "tools of yesteryear," 103 years of Northern New England Weather Center records, and the Carlton Felch diaries. Peggy fielded my call with a vigor that suggested things were pretty quiet around the Historical Collection office. The Fairbanks company's scales, she said, were ubiquitous from 1830 through the first half of the twentieth century. They were the Rolls-Royce of platform scales. When I told her Macdougall used a silk scale with a capacity of 300 pounds, she faxed me two pages of Fairbanks Silk Scales from a Macdougall-era catalogue.

"It combines great sensitiveness with the increase in capacity and platform," said the proud Fairbanks copy. "Very handsome in appearance." Indeed, as Macdougall had said, his scale was accurate to one-fifth of an ounce. I told Peggy the story of Macdougall and the dying consumptives, hoping she might have some nugget of Fairbanks scale minutiae that would explain the good doctor's three-quarter-ounce drop. She initially wondered whether the platform might have been inlaid, in which case, you couldn't have a cot sticking out over the sides of the platform without rendering the results, as she put it, "screwy." But the scale in the drawing had a standard suspension platform.

"That's about as far as I can take you," said Peggy. You could hear the disappointment in her voice. Peggy Pearl could tell me what the weather was on April 10, 1901, and she could tell me what Carlton Felch was up to that day, but she could not tell me whether Macdougall's cot contraption had somehow compromised the Fairbanks's renowned accuracy.

MACDOUGALL SEEMED AWARE of his study's weaknesses, and he encouraged others to try to extend and replicate his work. He wanted to do more trials himself, but was stymied by the earlier-cited frictions with officials. Overtures to "positioned and entrenched scientific authorities" at other facilities, he writes in a letter to R. Hodgson of the American Society for Psychical Research (ASPR), were met by rebuffs. His best bet would have been the ASPR itself. Indeed, the ASPR's scientific officer, Hereward Carrington, upon hearing of Macdougall's study, wrote at length and with great gusto, in an issue of the society's journal, about the possibility of outfitting a condemned prisoner in an airtight glass hood, and placing him, electric chair and all, onto a platform scale.

In the end, Macdougall resorted to weighing some dogs on a scale he set up in his barn. Owing to the difficulty of finding dogs dying from a disease that rendered them exhausted and motionless, he immobilized and then killed them via injection—fifteen dogs in all. Not one evinced a drop in weight as it died. Macdougall's spin on this rather striking batch of conflicting data was that of the churchgoing Christian: Animals don't have souls—or anyway, sayeth the Bible, not the eternal variety—and therefore we should expect this.

Not everyone in the soul-weighing business would agree with Macdougall there. Ten years after the *American Medicine* paper was published, a physics teacher at Los Angeles Polytechnic High School self-published a book called *The Physical Theory of the Soul*, which included a chapter detailing his adventures in mouse-soul-weighing. H. LaV. Twining, as he appears on the title page, didn't seem to like animals much, as we will see, but he gave them credit for arriving in this world with the same spiritual accoutrements as humans. "It is reasonable to conclude . . ." he wrote, "that all forms of life have accompanying souls . . . and that animals would form fit subjects . . . since they could be killed at will and under any chosen conditions, while human beings could not."

Over the following four pages, H. LaV. offs thirty mice on his scale, using every condition the Polytechnic supply closet would support. He suffocated them in test tubes melted shut by Bunsen burners (no weight loss). He smothered them in flasks sealed with rubber stoppers and flasks sealed with parafinned corks (no loss, and no loss). He gassed them in open flasks with cyanide pellets. Here at last he witnessed a loss: one to two milligrams "at its last kick." H. theorized that the poison had caused the mice to "perspire violently at death," and that the lost milligrams were evaporated mouse sweat. I did a little reading on the subject of cyanide poisoning, in the form of a paper by Dr. John M. Friedberg, which helped prompt the Ninth Circuit Court of Appeals to deem cyanide executions cruel and unusual. Death by cyanide does appear to be moderately aerobic—panic, retching, seizures, violent head extension, grimaces—though excessive perspiration isn't specifically mentioned as a symptom. Excessive salivation is, however. Perhaps some seizure-flung drool escaped the confines of the flask.

A colleague suggested that the lost weight might have been air expelled from the dying rodent's lungs. No slouch, H. decided to test for this, too, a process he describes, with his characteristic flair for insensitivity, like this: "A mouse was thrown into a tank of water." A test tube was slipped over the head of the drowning mouse to catch the air it expelled as it died, and this was weighed and found to be negligible.

H. quickly moved on to more pressing matters, such as the author of a book on Rosicrucian theory who had misstated some of H.'s conclusions and misspelled his name. "This is inexcusable," he crabbed. If humans, like mice, could be killed at will, I know just who would have been found in a tank with a test tube over his head. And along with him, the animal rights advocates who chastised Twining for his cruelty. (An account of his work had run in a local newspaper.) Here is Twining in defensive overdrive:

> Nearly every person who reads these lines has suffered more from the tooth ache, a thousand times more, than any of these mice did in dying. . . .
>
> Even though suffering does not take place, there is no reason why dumb animals should not have their share of suffering. Human beings during a lifetime are subjected to hours, days, and years of mental and physical anguish inflicted upon them through no fault of their own. . . .
>
> The human family lives on the products of death. . . . We eat either animal or vegetable food and in either case life is destroyed. . . . To kill low forms of life is just as bad, if killing be bad at all, as the killing of the higher forms of life, and there is no need of becoming hysterical over it.

The point to take away from H.'s work is that if you put a dying mouse in a sealed container—such that moisture, expelled breath, drool, et cetera, are trapped—its weight won't change. So H. LaV.'s work with mice is in line with Duncan Macdougall's dog findings, neither showing evidence of a departing soul.

IN 1998, DONALD Gilbert Carpenter published a whole book about soul-weighing (*Physically Weighing the Soul*). It's a long book but lightweight, as light as a soul, for it exists only in cyberspace, available by download at 1stBooks.com. According to Carpenter, the reason the dogs and the mice might have shown no weight loss at death is that their souls are so light they were below the scales' detection thresholds. Macdougall said his dog-weighing scale was accurate to one-sixteenth of an ounce (1.8 grams), but a dog's soul weighs less than 1.8 grams. How do we know this is the weight of a dog's soul? Because Donald Gilbert Carpenter has calculated it. (I *love* this guy!) Using Macdougall's findings for human beings—that the soul weighs about 20 grams—Carpenter calculated the ratio of soul-weight to body-weight-at-birth: 1 to 140. Applying this to a typical puppy birth weight, he deduced that the average dog soul weighs one gram—about half the 1.8-gram sensitivity of the scale. Same problem with Twining's mouse souls—too light to register. (But not Jesus' soul. The discarnate Jesus is calculated in Chapter 17 to weigh 364 grams—close to a pound!)

Elsewhere, Carpenter calculates the volume of the human soul—or Mac, as he prefers to call it, in honor of Duncan Macdougall. Here is how he came up with his volume amount. The smallest infant to survive at birth, he says, weighed ten ounces and had a volume of three-tenths of a quart. (I do not know the formula for calculating the volume of a premature baby; perhaps he hired H. LaV. Twining to throw one into a graduated cylinder and note how much water it displaced.) The volume of the baby's Mac would be identical to its birth volume because, quoting Carpenter, "if the volume of its . . . Mac had been bigger than that, it would have stuck out of the child's body." Once again, Jesus is the exception. His Mac had a volume of 5.25 quarts, meaning that half a quart's worth of excess soul stuck out of his body when he was born. Carpenter surmises the protruding material took the form of a glow, rather than the more pedestrian hump or goiter that leapt into my mind.

Carpenter points out that leprechauns have a volume similar to that of the human Mac. "This makes me suspect," he writes, "that Leprechauns . . . are most likely discarnate humans." This makes me, in turn, suspect that Donald Gilbert Carpenter is most likely not the staid scientist that his many equations and tables suggest. (Carpenter's bio says he knows more about materialistic research on the soul than anyone else alive, but it doesn't say what kind of degree he has or what he does for a living.)

Carpenter had not, at the time his book was published, undertaken any soul-weighing experiments himself, but he had some intriguing ideas. Rather than put dying people on scales, he thought it would be instructive to do the experiment with pregnant women, and look for a sudden weight gain around the moment the Mac enters the fetus—which he figures happens at forty-three days, when brain waves can first be detected. Carpenter outlines a variety of unique uses for pregnant women. On page seventy-seven he tells us, "An excellent way to de-haunt a house would be to make it the residence of newly fertilized women just prior to normal entry of the Mac into the fetus."

LEWIS E. HOLLANDER, JR., is a sheep rancher in Bend, Oregon. Sometime in 2000, Hollander, intrigued by Duncan Macdougall's work, became the second man in history to set up a soul-weighing operation in his barn. He rigged a seven-by-three-foot platform to a Toledo model 8132 electronic digital indicator, a quartet of load cells, and a computer. His subjects were eight sheep, three lambs, and a goat, all of which were sedated and then euthanized, and all of which, he assures us, were headed that direction anyway. The animals were wrapped in plastic to, as he put it, contain any voiding. This was important because (a) voided material might drip off the weighing surface, creating a spurious weight loss, and (b) you try getting sheep urine out of your load cells.

Though his goals march lockstep with Twining's, the similarities stop there. Hollander is a kindly, soft-spoken guy, and he genuinely likes sheep. "They're easy to deal with," he told me, "and there's a whole lot of warm things about them." Hollander did not relish extinguishing that warmth. "I don't know if you've ever killed anything, but it's a very traumatic thing to do. To sit and watch this animal . . ." That was why his subject pool was limited to twelve. (He had actually contacted local doctors about the possibility of weighing end-game hospital patients, but ethical issues proved insurmountable.)

Here is the odd thing. All the sheep Hollander tested showed a temporary weight *gain* at death—most between 30 and 200 grams. One notable ewe put on 780 grams: nearly two pounds (or 37 Macs, or two discarnate Jesuses). The gain lasted from one to six seconds and then it disappeared. The three lambs did not, however, gain any weight, and neither did the goat. I called Hollander and asked him what he thought this meant.

"I haven't the faintest idea," he said sensibly. He acknowledged the possibility that the weight gain was an artifact of his equipment malfunctioning, but his instinct was that the blip was real. "If you were there at the time, you could see the whole scenario coming together and you could see this moment . . . It's weird. There was something happening there."

What could possibly have been happening? Hollander's feeling is that the changes had to do with what he called a portal to the beyond. "I think that at the moment of death that little window opens up. I think that maybe we're all connected to something bigger than we are."

I'd buy that notion. But why would opening it make an organism gain weight? Is the window a Dairy Queen drive-through? Carpenter, who has a section on Hollander's experiment in his book, theorized that the added weight was that of visiting Macs. (By his calculations, a seventy-kilogram human being holds a standing-room capacity of 280 Macs—one to direct the body and other optional ones to "perform less clear roles.") He noted that the sheep's mysterious weight gains got progressively larger. "It was as if each sequential death attracted more Macs to the scene," he wrote, though he couldn't say why, or why they departed again after six seconds, or what they have against goats.

YOU MAY BE relieved to hear that my next guest does not believe in leprechauns. He has an M.D. from Stanford and an undergraduate degree from Yale in chemical engineering, with a special interest in thermodynamics and information theory.

Nor does he have a friendly nickname for the soul. He has the opposite of a friendly nickname. He calls it "the (obligatory) negative entropy (i.e., energy/weight equivalent) that is necessary to allow for the nonequilibrium meta-stable physical 'quasi-steady-state' of a living/conscious biological system." And he has a plan to weigh it.

Gerry Nahum is a professor at the Duke University School of Medicine who works in an atmospheric old building called Baker House. The building holds an unlikely mix of what I'm guessing is runover from the rest of the medical center. Nahum shares the second floor with the Brain Tumor Center, Dr. C. H. Livengood, Pastoral Services, and the jolly-sounding Endocrine Fellows Office. He himself teaches obstetrics and gynecology. When I first learned this, I thought that perhaps Nahum had been scheming with Donald Gilbert Carpenter, and that any day now, the two of them would be ushering forty-three-days-pregnant women onto extraordinarily sensitive Duke University scales and watching the readout for the arrival of the Mac/(obligatory) negative entropy. Hardly.

Nahum is leaning back in his desk chair, fiddling with his tie, listening to me sputter about what it is I want to know. The tie is patterned with a university logo, very much in keeping with the decor, which is: thirty-one framed diplomas, degrees, and award certificates.

I've just described to Nahum the experiments of Duncan Macdougall, hoping to get his professional opinion regarding what might have caused the mysterious weight losses. A flicker of worry crosses Nahum's brow. Before I arrived, we exchanged a few e-mails, but I failed to fully prepare him for the depths of my ignorance. My ignorance is not merely deep, it is broad; it is a vast ocean that takes in chemistry, physics, information theory, thermodynamics, all the many things a modern soul theorist must know. Nahum pronounces

Macdougall's experiment "silly." He says you'd need not just a scale, but a completely isolated system.

This system—which Nahum would very much like to build—would be a sort of box, a box completely isolated from the surrounding environment. The box sits atop a mind-blowingly sensitive scale, and all around it are arrays of electromagnetic energy detectors. These detectors measure all the different types of known radiant energy (as opposed to informational, or "soul" energy, for which there are no detectors) that might leave the box. Now let's say there's an organism in the box—a paramecium or a wombat or John Tesh; it doesn't matter. And that organism dies inside the box.* If the electromagnetic detectors detect energy leaving the box, there should be a corresponding change in weight. Why? Because of the laws of physics: There is always a weight loss associated with an energy loss. I'm not talking about the listless feeling that besets the overambitious dieter. I'm talking about $E = mc^2$. If the energy changes, then the mass (which is proportionate to weight) must change—in, you know, a teensy, tiny, infinitesimal physics lab way. So if the mass lost when the organism dies is more than what would be expected based on the energy

*Credit for the original seal-a-soul-in-a-box experimental format must go to Frederick II, the thirteenth-century King of Sicily and Emperor of the Holy Roman Empire. In the diaries of the king's sometime chronicler, the Franciscan monk Salimbene, there is a description of Frederick shutting up "a man alive in a cask until he died therein, wishing thereby to show that the soul perished utterly." Though Frederick is to be credited for his precocious enthusiasm for scientific method, the cruelty of his experiments invariably outweighed their scientific merit. To wit, the time he "fed two men most excellently at dinner, one of whom he sent forthwith to sleep, and the other to hunt; and that same evening he caused them to be disemboweled in his presence, wishing to know which had digested the better" (the sleeper). At least that one makes some sense.

change, then something's leaving the box in a way that can't be accounted for. That something being, perhaps, the soul, or consciousness, heading out to some higher dimension—Lew Hollander's place beyond the open window.

Theorists like Nahum think of the consciousness as information content. And information, to a quantum physicist anyway, has an accepted energy equivalent. And thus a (very very very light) weight. "The change in the heat that has to be liberated per bit of information lost is about three times ten to the minus-twenty-one joules," Nahum says.

I must have made some sort of face. "I'm making this as simple as I can," Nahum says. When you're as brainy as Gerry Nahum is, you lose sight of just how ignorant the rest of us are. Earlier on in our talk, he prefaced the line "Quite a few people look at microtubules as what can be considered almost like an abacus for molecular calculation at a subcellular level" with the phrase "As I'm sure you're aware."

For the sake of not completely derailing the explanation, we're going to accept that the energy lost when one unit of consciousness information is destroyed has been determined by pedigreed physicists to be three times ten to the minus-twenty-one joules, and let Dr. Nahum continue. "And if you use the mass-energy equivalent equation"—the Einstein thing—"then you can say, 'Well, if that's true, then that has to represent three times ten to the minus-thirty-eight kilograms.'" So the weight of one bit—the basic unit of information, the stuff that makes up human consciousness—is a billionth billionth billionth of a billionth of a kilogram. "It's very small," says Nahum, and that I understand.

But how many bits are there in a consciousness? Or in one thought? When I think to myself, *Is this man blowing smoke up my microtubules?* how many bits are involved? Not known. "Is a thought one billion bits?" says Nahum. "Ten billion? We don't know. When we look at consciousness, what is embodied in that? How many bits? We don't know." In a way, it doesn't matter. What's important for our does-the-soul-exist purposes is that changes can be detected. The energy loss created by a soul heading out the window can, in theory, be detected as a weight loss.

The Fairbanks company does not make a scale for Gerry Nahum's purposes. Does anyone? Possibly. Scales have traveled a surprising distance since Macdougall's day. There are scales that easily and accurately measure micrograms, a microgram being a millionth of a gram. Measuring a billionth of a gram—a nanogram—is also possible, though costly. "What about a picogram?" muses Nahum. That's a trillionth of a gram—10 to the minus-15 kilograms. "Can we measure that? Yeah, we can. Remember I gave you the figure ten to the minus-thirty-eight kilograms?" I remember: in the discussion of the weight of one bit of information. "I've just told you I can measure fifteen orders of magnitude of that. The question is, can I measure the next twenty?" Maybe he doesn't have to. Assuming the consciousness is made up of a vast number of bits, maybe he can get away with the picogram scale.

Nahum says the electromagnetic field arrays around the box are more problematic than the scale. None of these detectors operate over the entire electromagnetic spectrum, so Nahum would have to overlap and improvise. Despite this, he thinks it could be done.

But what if the soul—the residual energy/information that doesn't register on our electromagnetic energy detectors—doesn't go somewhere else, but just, you know, snuffs out? Ceases to exist? That has always been my own depressing, humdrum assumption regarding death. No can be, says Nahum. Standing in the way is the First Law of Thermodynamics: Energy is neither created nor destroyed. It has to go somewhere. Nahum says he became convinced that this applied to the consciousness when he was five years old. Around the age you and I were puzzling out the intricacies of the shoelace, Nahum was "thinking about how it had to be conservative, that there's no way out." Nahum swivels to face me. "The question then becomes, Where does it go? The question is not, Is it there? It's there."

We sit quietly for a minute, allowing the guest to absorb this rather dense helping of quantum theory. In a corner of the ceiling, a fluorescent light flickers and goes out. Applying the First Law of Thermodynamics, we know that elsewhere in the universe, an unattractive though cost-efficient glow has just appeared.

Though Gerry Nahum has long been consumed by matters of the soul, he is not a religious man. However, he has had some interesting encounters with the Catholic Church. "I approached them, naively, years ago. To get funding. I outlined it like I just did for you." I picture the bishops in their high-backed chairs, Nahum tucking "Your Excellency" here and there in his warp-speed, single-spaced prose.

The monsignors didn't understand the specifics of Nahum's proposal, but they understood enough to know that it made them nervous. "They have a system of belief where they *know* what the answer is. They don't need quote-unquote proof. And if [the results don't] agree with what they know, it's a disaster. They don't want to take that risk." After Nahum's first audience, he was invited back. Now the mood had grown solemn. Outside experts had been called in, theologians with backgrounds in cosmology and physics. Not only did they not offer to fund Nahum's project, they did their best to talk him out of it. They spoke of a "divine design" for the division of the worlds, and tried to make the case that Nahum's experiment threatened a breach of that division. The consequences, they warned, could be dire and unfathomable. "They envisioned that there was a potential for opening a dark 'schism' that might unleash some type of heretofore unknown 'power' into our traditionally protected world." The window metaphor made an appearance. Nahum was accused of trying to "open a window that might not be closed after opening."

But the window presumably opens by itself, whenever something dies. Why would they think Nahum is trying to pry it open? Why would his experiment keep the window from closing? Why can't souls use doors like the rest of us?

At the last meeting, the bishops tried to open a window of the stained-glass variety. "They suggested I might seriously consider converting to Catholicism, so that I'd get over the whole idea. In the end, I had to take a 'just kidding' stance and essentially feign that I had no further interest in pursuing it."

These days, Nahum trolls for funding at physics departments and institutes like the University of Arizona's Human Energy Systems Laboratory. He goes to science-of-consciousness and quantum theory gatherings whenever he can, hoping to hook up with potential partners. It is going slowly. "Most people don't listen nearly as long as you have." Yes, I say, but they probably listen better. Yes and no, replies Nahum: "It's a multidisciplinary idea, so it's a tough sell. The engineering and information specialists know nothing about biology. The physicians and the biologists and the neuroscientists know nothing about information theory. And none of them know anything about cosmology or . . . the physics of multidimensional universes. They're very smart people, but they don't have the breadth of background to incorporate it all into one." Nahum is like the discombobulated animals in those children's books where the pages are split into thirds and the ostrich has a kangaroo's legs and the hippo is part giraffe. He's a little of everything, and there's no one for him to play with.

The closest he's come to a soul mate is Patrick Lui, who manages R&D collaborations at the Stanford Linear Accelerator Center and studied thermodynamics in graduate school. I spoke to Lui after I got back from Duke. Lui told me he tried to get other physicists at Stanford, including the former chair of theoretical physics, to "think along with Gerry." Both Lui and his colleagues felt that although the concepts were valid and the project made sense intellectually, the experiment would be difficult or even impossible because of the challenges of measuring such extremely small amounts of energy. "That doesn't mean that one should not pursue this kind of work," Lui was quick to add. "This is a curveball, but nonetheless it is a real ball."

Nahum's idea is also a tough sell among nonacademics. Because, as he puts it, "People either think they already know the answer and don't want any external validation, or they think it's impossible to know the answer. They don't have enough of a background to understand that they *can* know."

And then there are budget constraints. Nahum estimates he'd need at least $100,000 in funding. "People have said to me, 'We're going to take this up the ladder, see what we can do.' But it's not a mainstream, high-priority enough idea that anyone says, 'Here, let's commit the money.'"

What about the physics department here at Duke? "I get blank stares."

I begin to feel sad for this misunderstood man with his grand and misunderstood—or just not understood—vision.

"Does your wife understand your project?"

"Ex-wife. Not even a little."

Nahum takes a phone call from a business partner named Al. "You're wrong, Al!" he is yelling good-naturedly into the phone. "Al . . . AL! You're all wet, Al!"

I put away my gauzy pastoral of the lonely philosopher.

Gerry Nahum is a tall, charming, pedigreed gynecologist with healthy self-esteem. One day he'll get the support he needs to carry out his plan, and possibly the respect of the Duke physicists, and maybe even a wife who knows quantum theory. I hope so.

IT IS 2 P.M. before Nahum's stomach makes itself heard over his brain and we break for lunch. With the equations put away and at least a few picograms of Nahum's informational content devoted to his ravioli, I feel more comfortable asking the dimwit questions I've wanted to ask all morning.

What do you think it would feel like to be a free-floating soul, a fart of energy in some god-knows-where-or-what dimension? Nahum makes the analogy of the computer: Your basic core of consciousness, he imagines, would be like the operating system. On top of that you have various overlays: word processing and spreadsheet programs and such if you're a computer; if you're a human being, perception, language, reason, memory. When you die and the brain shuts down, the overlays fritz out. You're left with the operating system: a sort of a primitive, free-floating awareness. Nahum imagines existence would be "like what it is for us now, minus all the superficial trappings."

It's that minus-all-the-trappings bit that gets me. If you can't think in words or see or hear, what are you like? Coma victim? Lichen? Nahum shrugs. It's just an analogy, just a guess. I posed this question to Lui later in the week. He was dubious about the possibility of the informational content of a person's consciousness leaving the body in any sort of organized form. "Decay heat is not ordered information," he said. Meaning, I think, that the blip of energy that was your personality may indeed continue to exist after you die, but not in the form of your personality. Not in the form of something you can be or use.

I later relayed to Nahum what Lui said and asked him to comment. “Remember,” I wrote, “in replying to me, pretend you are talking to a seventh-grader.” Nahum disagreed with Lui. His reply ran to a thousand words and would have been understandable to any seventh-grader familiar with Kant, Locke, negentropy as the measure of nonrandomness, and the Enigma encryption machine. Here is the part I understood: “This energy is freely malleable in terms of the physical form it might take . . . and it is not necessarily the case that any one of them would be ‘preferred.’”

Well, “preferred” in the sense that it would be more fun to be a spirit that can think thoughts and remember memories than to be, say, a black hole or a piece of static electricity. But I decided to leave it be.

Nahum orders bananas Napoleon for dessert, just one more way we’ll never understand each other. We’re back to talking about the box, the system. I realize I forgot to ask him what kind of organism he’s planning to put inside. The dessert arrives, a massive custard download held vertical by wafer shelving.

“So what goes in it?” I’d been assuming a lab mouse.

“Banana pudding, mostly.”

When I get back home and I look at Nahum’s twenty-five-page “Proposal for Testing the Energetics of Consciousness and Its Physical Foundation,” I will picture a plate of banana pudding in a box.

Theoretically, Nahum could sacrifice anything from a bacterium on up. He is leaning toward leeches. “I worked with leeches for a long time. They’re slimy and they latch on to you. They’re a very awful organism. I hated those things!” The couple at the table beside us turn to look at the man who hates leeches.

Last question: What does he think the result will be? Does free-floating consciousness energy exist? "My bias is that it does exist," says Nahum. "But I would never say that I know that." He puts down his spoon. "Until I prove it."

4

The Vienna Sausage Affair

And other dubious highlights of the ongoing effort to see the soul

THE YEAR WAS 1911. Duncan Macdougall, feeling fairly certain he had proved the existence of the soul (by weighing it), now was determined to see it. He wanted, he said in a newspaper article from that year, to know what color* the soul was and how large it was in relation to the body. He wanted to know what route it took on the

*Anecdotal data on this matter comes from a former nurses' aide named Juli Pankow, who e-mailed me regarding her observation of what she took to be the soul of a dying nursing home patient. The room was dark. She had just heard the woman's death rattle. "There was a greenish-purple very very faint cloud or haze right above the chest." From a Google search, I learned that exploded barium can appear as a greenish-purple cloud, though the cloud in question was linked to a NASA project, not post-barium-enema gas, so who knows.

way out. Did it emanate from the heart or the top of the head, or did it perhaps escape "from the lips," like a yawn or a cartoon character's speech? Ignoring the growing disapproval of hospital and asylum boards, Macdougall recruited yet another batch of fading consumptives.

This time around, the dying men lay on an ordinary bed, in a darkened room. At "the supreme moment," as Macdougall put it, he aimed a strong beam of light along the length of the patient's body. First he used a swath of white light, and then he experimented with the individual colors of the spectrum, using a long glass prism to separate the bands of light. He tried it with the spectrum horizontal and then with it vertical. He detected nothing.

Macdougall concluded that the soul's index of refraction is zero, and since all substances except "the ether of space" give some refraction of light, the soul was therefore made of ether. A word about ether. In pre-Einstein days, ether was an accepted concept in physics. It was thought to be the necessary medium for the transmission of electromagnetic waves (light, sound, radio, et cetera). Ether was invisible and undetectable, and it permeated all forms of matter, from man to ottoman.

It was also assumed to be weightless, which threw a hitch into Macdougall's soul-as-ether idea. Unwilling to abandon his theory, Macdougall went public in 1914 with an "astounding theory," as one headline put it, about how both ether and souls in fact were subject to the laws of gravity. Because if they weren't, he reasoned, it would follow that "ever since human beings began to die upon the earth, the complex pathway of the earth around the sun in space . . . [would be] littered with these . . . nongravitative spiritual principles." In other words, if gravity didn't hold dead souls here on earth, they'd drift away into space, relegated to an eternity among the derelict satellites and NASA detritus. "Can we construct a conception of an orderly future life out of such conditions?"

Equally untenable to Macdougall was the thought of one's soul being separated from one's earthly remains by millions of miles. "Between the time of man's death and the time of the burial of the body, an average of three days, as a spiritual principle he would be separated from his body and the place he died by a distance of nine million miles, the distance the earth would travel in three days' time." To a man who spent his adult life five thousand miles away from his parents and siblings in Scotland, you can imagine it was a gloomy proposition.

Macdougall envisioned a giant "earth-accompanying globe of ether . . . above the storm zone," a sort of floating reunion hall for "beings constituted entirely different from us, who are yet subject to gravitation." Perhaps singed by his colleagues' earlier scorn, the doctor published neither his refraction experiments nor his ether-heaven theory in *American Medicine*. The only description I have was part of a 1914 *Boston Sunday Post* article entitled, quite marvelously, "Heaven Is Perhaps Just Outside Earth."

And that was the last to be heard from Dr. Duncan Macdougall. Six years later, he took to his bed with cancer, wrote one last awful poem ("I had a bout with Death / We strove through night and day"), and departed for the great globe of ether. His wife Mary, aforementioned battleaxe, lived on for another thirty-five years. Depending on whether Macdougall was right or wrong about gravity's hold on souls, this could mean that when the missus' soul finally shed its earthly shell, Duncan's own soul would be thirty-eight billion miles away. To every cloud, a silver lining.

Around the time Macdougall went public with his disappointing light-ray project, a University of Pennsylvania physicist named Arthur W. Goodspeed trumped him by announcing plans to reveal the human soul by way of the amazing new roentgen ray (now called X-ray), named in 1895 for its discoverer, Wilhelm Roentgen. (Goodspeed had inadvertently discovered the rays some time before Roentgen had, but failed to recognize the import of what he'd done and watched his rival Roentgen become a household name, albeit a mispronounced one, while he faded to obscurity.)

X-rays are today a humble diagnostic tool but in their infancy were considered nothing short of miraculous. Nan Knight, director of the archives at the history center of the American College of Radiology, told me that Thomas Edison, who seems to have invented publicity along with everything else he invented, at one point announced a public demonstration in which he would take an X-ray photograph of the living brain,* showing actual thoughts as they darted here and there. Within a year after the ray's discovery, Parisian hucksters were selling tickets to sideshows purporting to show ghosts captured as X-ray images. In 1896, a New York newspaper reported that over at Columbia's College of Physicians and Surgeons, X-rays were being used to project anatomical diagrams directly into the brains of students, "making a much more enduring impression." Somewhere along the way, a rumor surfaced to the effect that opera glasses could be outfitted with X-rays, considerably upping the appeal of a night at the opera for many a bored spouse. So thoroughly taken was the public by this story that on February 19, 1896, a New Jersey assembleyman introduced a bill specifically—and to the

---

*In reality, an X-ray of the head could not show the brain, because the skull blocks the rays. What appeared to be an X-ray of the folds and convolutions of a human brain inside a skull—an image that circulated widely in 1896—was in fact an X-ray of artfully arranged cat intestines.

great derision of his peers—prohibiting the use of X-rays in opera glasses.

Temple University Urban Archives in Philadelphia has a file of news clippings on Arthur Goodspeed. His soul project made the *New York Times* on July 24, 1911, under the headline "As to Picturing the Soul." The article quotes none other than Duncan Macdougall, denouncing the project's worth. Though he tempered his skepticism at the end of the piece by admitting that "at the moment of death the soul substance might become so agitated as to reduce the obstruction that the bone of the skull offers ordinarily to the Roentgen ray and might therefore be shown on the plate as a lighter spot." A separate article mentioned that Goodspeed was to be assisted by "his Roentgen ray expert Dr. Snook." Though biographical material on Dr. Snook* makes no mention of soul X-rays, history has bestowed at least one honor upon the man. He is known to this day for the Snook tube, an obsolete glass-globed cathode tube that resembles a hummingbird feeder.

If Goodspeed wrote a paper on his soul X-ray project, the archives didn't have a copy. Nan Knight wondered whether his mention of it might have been a joke. Possibly, but I don't think so. Not only did Goodspeed's bio list him as the vice president of the American Roentgen Ray Society, but it named philosophy as his chief interest (with "trotting horses" a close second). He was the secretary of the American Philosophical Society for thirty years. Since he ran the physics department's Randall-Morgan Laboratory, he would have had both the equipment and the budget for such an undertaking. Plus, we have the requisite untimely loss of a loved one, which so often

*To those who find humor in this poor man's name, I have this to say: His full name was Homer Clyde Snook.

sparks an interest in the hereafter among otherwise orthodox scientists. Goodspeed's twenty-eight-year-old son died in a parachuting accident.

For an even more creative approach to soul-viewing, we have Hereward Carrington, Ph.D., the founder, in 1920, of the Psychic Laboratory and the undisputed gizmo geek of the paranormal set. In his 1930 book *The Story of Psychic Science*, Carrington sets forth his idea for a machine to reveal the shape of the soul. The description takes two pages, beginning with "arrange a small box as to imprison some animal—a dog, cat or small monkey" and ending with "therefore, when condensation occurs, the resulting line will *outline the form of the astral body*." Along the way, we make stops for hermetic seals, piped-in anesthetic, dust-free air, ionization rays, an air pump, and no doubt several Snook tubes. The book contains a half dozen photographs of the dashing Carrington, his Gregory Peckish hair swept back from his forehead and a scowl of concentration on his face, checking the readings on the dials of his latest gadget. The photograph invariably includes an attractive young woman, sometimes hooked up to the machine, other times merely gazing rapturously at Dr. Carrington. I had a crush on Dr. Carrington, too, until I saw some of his later titles, which include *The Hygienic Life* and *Fasting for Health and Long Life*.

Carrington never built his monkey box, so the honors for most elaborate soul-manifesting device go to a pair of Dutch physicists, J. L. W. P. Matla and G. J. Zaalberg van Zelst. Matla believed himself to be in contact with an entity who spelled out communications letter by letter on a Ouija board. (Hopefully the question "What is my full name and that of my partner?" was never posed.) The entity informed Matla that the human spirit survives death to become a gaseous body called *homme-force* (meaning "man-force"; this was the French edition, which was all I could find). Matla reasoned that if *homme-force* were truly a gas, it must obey the laws of physics, and thus its existence might be proved scientifically. The entity, which appeared to have an engineering background, not only agreed with this, but provided detailed instructions for the building of a device for the task. The transcript of séance #36 reads, in part: "Construct two cardboard cylinders that are impenetrable to air. Length 50 cm. Diameter, 25 cm. . . ." The idea was that the gaseous entity would pass into the cylinders and make its presence known by expanding and contracting on demand. The displaced air would then trigger the rise and fall of a drop of grain alcohol in a glass tube. Which it did, at least to the satisfaction of Matla. The pair sallied forth with more and more elaborate calculations. Their book *Le Mystère de la Mort* is full of lines like "*Le volume de la masse de l'homme-force est de 36.70 m.M*^{3}" and "*Le volume du gaz déplace 279.169 c.M*^{3} *d'air.*" It's hard to say where is the bigger hubris, in their convictions or in the arrogance of carrying them to a third decimal point.

Hereward Carrington read the book and, being Hereward Carrington, couldn't resist building a set of Matla cylinders of his own. His model was accessorized with a bell, rigged to ring whenever the alcohol drop moved. Carrington would assemble a group of observers in a room with the device, ask for quiet, and then announce loudly, as though man-force had hearing trouble: "If there be any force present capable of entering the cylinder and thus displacing air, will it please do so?" Sometimes the alcohol drop moved, but not necessarily in conjunction with a request to do so. Often repeated requests were made and nothing happened. Then Carrington would give up and leave the room, only to hear the bell ring. It was as though the entity were intentionally making Carrington look like a boob in front of whichever fetching young lab assistant was at his side. Carrington spent an entire year messing about with the cylinders. In the end, he concluded that temperature changes and coincidence accounted for Matla's results.

The amazing thing about men like Matla and Carrington is that they weren't perceived as fringe scientists. For a significant span of years, paranormal research was an accepted undertaking among respectable scientists. As proof, I offer an article that ran in the July 30, 1921, *Lancet*, then and now one of the most respected medical journals in the world. A Dr. Charles Russ writes that he has proved the existence of an unknown "force or ray" that "emerges from the human eye." Russ built a tabletop device to demonstrate this force, which is illustrated in the article. Volunteers were asked to stare into a sealed box at a copper-wire solenoid suspended from a string and held steady by magnetic force. If they stared at the left side of it, he claimed, the solenoid would be pushed clockwise; right-side staring set it twisting the other way. He posited that the mystery force "disturbs the electrostatic state of the enclosed system" and claimed that five physicists with London's Royal Society were unable to find any electrical or mechanical fallacy that could account for the effects. Meanwhile, it's business as usual on the facing page, with dysentery expert W. S. Dawson holding forth on fecal sampling—whether it is preferable to "introduce the swab" into the rectum or to take a specimen directly from "the motion."*

---

*I'm trying to work out how this makes sense as a noun meaning "the product of a bowel movement." This is not Dawson's personal euphemistic misstep; the usage persists in medical writing today. Should you have had the misfortune of visiting a web page called the Constipation Page, you will have seen the phrase, "the motion or stool is very dry or hard." Perhaps this is why the term "motion pictures" was replaced by "movies." Now that I see it on the page, "movie" would have been a far better BM euphemism than "motion." *I'd love to chat, but I need to make a movie.*

As for Dr. Russ's machine, I can't explain what was going on; however, given that no one has since replicated the study, I suspect it was a load of motion.

Dr. Russ was but one of a long parade of scientists who were convinced that the soul was a sort of life force that—if not actually photographable or locatable as an entity—could be detected and proved indirectly, by way of its emanations. Often these emanations were captured as coronal glows or lightninglike rays around the perimeters of living objects placed in contact with a photosensitive surface. Bear in mind, photography was in its infancy; its processes were still poorly understood. While some of the manipulations were outright fraud, most were the result of sincere, if misguided, efforts.

For instance, the effluviograph. In 1897, a team of scientists from the Société de Biologie in Paris demonstrated a new technique in which a halo could be made to appear on a photographic plate where a subject had held his finger. The men assumed they had discovered a way to capture od rays. Od force had been making the rounds as the latest form of life force, having bumped aside Franz Mesmer's animal magnetism. In reality, what the French biologists had discovered was the effects of heat on photographic developer. As proof, chemist and debunker Emil Jacobsen produced an effluviograph that looked identical to the Parisian fingertip effluviographs but was made by touching the plate with the rounded end of a heated glass test tube.

Jacobsen's next undertaking was to unseat the electrograph. In *Beyond Light and Shadow*, Rolf Krauss's admirable history of paranormal photography, you will find a reproduction of an 1898 photograph entitled: "Electrograph of the antipathy between two Vienna sausages." The photo was part of Jacobsen's paper debunking electrographs, which do no more than document the light from an electrical discharge given off by an object—living, dead, or inert—that has been hooked up to an induction machine and placed in contact with a photosensitive surface. Using this technique, a member of the Imperial Institute for Experimental Medicine in St. Petersburg had claimed to be able not only to photograph the life force, but to use these images to diagnose illnesses and emotional states. The scientist paraded a photograph of two adversaries holding their fingertips close together—the inspiration for Jacobsen's satirical Vienna sausage series—such that their sparks of nerve force could be seen to keep their distance, whereas nerve force among friends could be seen to commingle.

But electrographs did not go away. They were resurrected in the 1950s by a Russian couple named Semyon and Valentina Kirlian. Kirlian photography eventually became known as aura photography and to this day maintains a presence at psychic fairs. Interestingly, the Kirlians—despite sounding like a robe-wearing doomsday cult—never claimed that they'd found a way to photograph the soul or the astral body or what have you. Both the Kirlians and subsequent Kirlian-inspired parapsychologists—including a pair at UCLA—noticed a lot of variation in people's "auras," both from subject to subject and from hour to hour with the same subject. For this reason, they surmised that the photographs might prove a useful diagnostic tool. To figure out what it was they were diagnosing, the UCLA team hooked subjects up to induction machines and exposed them to all manner of physical and emotional stimuli. They were given drugs, plied with alcohol, asked to meditate, pissed off, frightened, and hypnotized. Although differences were observed from photo to photo, no useful patterns were found. Science stepped away from aura-reading, and the New Age—aided by the invention of color photography—stepped in.

The most famous Kirlian photograph is of a leaf with a missing tip, taken by Russian parapsychologists circa 1970 and presented as evidence of a sort of astral body made up of "bioplasma substance." A glowing outline of the entire leaf, including missing tip, appears in the photo. I'm too undone by the concept of trees having astral bodies to register much of an opinion on the Russians' work, though I can tell you that at the time Krauss's book went to press, no one claimed to have reliably reproduced the phenomenon.

You will not be surprised to learn that Hereward Carrington did some aura research of his own. (This was the 1920s, long before the Kirlians had brought photography to bear on the subject.) In *The Story of Psychic Science*, he mentions that his "experiments with Negroes" seemed to suggest that the aura was either a subjective impression or an optical effect. I could find no details of Carrington's aura experiments, and that is probably for the best.

And then came ectoplasm—Macdougall's soul substance at last revealed. Ectoplasm debuted in 1914, in a series of bizarre photographs in an equally bizarre but briskly selling book called *Phenomena of Materialization*. By 1922, it was the stuff of major newspaper headlines. I can say with the steely confidence of someone who has reported on Vienna sausage emanations that no stranger episode ever beset the march of science.

Hard to Swallow

The giddy, revolting heyday of ectoplasm

THE LIBRARY AT Cambridge University has its very own admissions office. This is where you are sent should you be so daft as to try to walk in without a Cambridge ID. I'm waiting in the hallway outside, to apply for one-day admittance. Specifically, I'm trying to get into the hallowed Cambridge Manuscripts Reading Room, overseen by the Keeper of Manuscripts and University Archives, whom I picture standing guard at the door in ankle-length robes with a massive key on a chain around his neck. I'm actually nervous about getting in.

While I wait, I read about the sacred texts on exhibit in the lobby. *Amongst the many Buddhist works in the Cambridge University Library is this very important Sanskrit palm leaf manuscript, about 1,000 years old. . . . Cambridge University has one of the most important collections of Buddhist Sanskrit manuscripts in the world.*

Meanwhile, yours truly is here for archive item SPR 197.1.6: Alleged Ectoplasm.

Ectoplasm lived during the table-tipping, spirit-communing, strange-goings-on-in-the-dark heyday of spiritualism. It was claimed to be a physical manifestation of spirit energy, something that certain mediums—called "materializing" mediums—exuded in a state of trance. "This stuff seems to diffuse through the tissues of the [medium] like a gas, and emerges through the orifices because it passes more freely through the mucus membrane than through the skin," wrote Arthur Findlay, founder of the Arthur Findlay College for mediumship and other spiritualist pursuits. The spiritualists described ectoplasm as a link between life and afterlife, a mixture of matter and ether, physical and yet spiritual, a "swirling, shining substance" that unfortunately photographed very much like cheesecloth.

The original ectoplasmic medium was Eva C., whose emanations drew the attentions of French surgeon and medical researcher Charles Richet. Richet was the discoverer of human thermoregulation and cutaneous transpiration, a pioneer in the treatment of tuberculosis, a recipient of the Nobel Prize for his work on anaphylactic shock, and the author of *Gastric Juice in Man and Animals* (can't have a slam dunk every time). That a man of his stature spoke for the authenticity of ectoplasm made it difficult to dismiss. As did spiritualism's roster of scientists, statesmen, and literary luminaries: William James, William Butler Yeats, Sir Arthur Conan Doyle, physicist Sir Oliver Lodge, chemist Sir William Crookes (inventor of the vacuum tube and sufferer of ridicule for his pronouncement that the luminous green gas inside his invention was ectoplasm), two prime ministers, and Queen Victoria.

Spiritualism, in a nutshell, is a religious movement devoted to communicating (via mediums) with those who have died and to proving to others, via séances and other mediumistic demonstrations, that it is possible to do so. Death is viewed not as an end to life, but merely a different phase, a changing of address and scenery. Heaven—or Summerland, as the spiritualists used to call it—was no longer an abstract but a place you could put a call in to. Spiritualism was founded in 1848, by the elder sister of two bored preteens, Margaret and Kate Fox, who took to soliciting mysterious spirit "rappings" at their farmhouse in Hydesville, New York. The noises stirred the imaginations of local townsfolk and the entrepreneurial spirit of their sister, who was soon inviting strangers to the house to observe the proceedings for a modest fee. Within months the three sisters were on a nationwide tour, and spiritualism was off and running. It spread steadily and traveled overseas, peaking in the aftermath of World War I, which left millions of American and European families grieving for lost sons and sadly vulnerable to the promise of contacting them in the afterlife. Though spiritualism's ranks have dwindled since, it retains a presence in the United States and, more prominently, England.

In 1989, possibly to its great embarrassment, Cambridge University acquired the archives of the Society for Psychical Research,* the preeminent investigators of the early mediums'

---

*I am an unabashed fan of the SPR (which has been around since 1882) and in particular its quarterly journal. Here are peer-reviewed articles addressing in all seriousness the likes of wart-charming and talking mongooses. Here are time-domain analyses of table rappings and field studies of healers' effects on lettuce seed germination ("Figure 2: the healer 'enhances' the seeds, mimicked by the control healer"). I take it as nothing beyond happy coincidence that the SPR membership roster has at one time or another included a Mrs. H. G. Nutter, a Harry Wack, and a Mrs. Roy Batty.

claims and feats. Should you wish to view, say, the file labeled "Merton, Mrs.: Investigation of 'The Flying Armchair'" or "Gramophone Records—Alleged Trance Speech of Banta," a young, madonna-skinned Manuscripts Room page will find it and place it in front of you with the same reverence and respect he accords items of the Royal Greenwich Observatory Archives or the seed specimens of Sir Charles Darwin.

Unlike the flying armchair, which officers of the SPR swiftly dismissed as hokum, ectoplasm was the subject of elaborate and stone-serious scientific inquiry for more than two decades. *Scientific American* sponsored an investigation of materializing mediums that was covered in four consecutive issues during 1924. In 1922, the elite Sorbonne University in Paris assigned a team of scientists to sit in on fifteen séances with Eva C., with the specific goal of testing the authenticity of her ectoplasm. (It flunked.) The September 1921 *Popular Science Monthly* observed that ectoplasm "can assume the shape of a hand or a face or even a whole figure" and is "curiously like human skin in cellular structure." In 1922, Harvard University graduate student S. F. Damon was featured in the *New York Times* for his belief that ectoplasm was the elusive "first matter" of the ancient alchemists. The *Times* index for the years 1920 to 1925 includes more than a dozen entries under "ectoplasm," ranging from straight-faced coverage of research to more farcical forays, such as "Man Bites a Ghost and Upsets Seance" ("Gallagher actually got a mouthful of ectoplasm. . . ."). Yet you look at any one of the hundreds of photographs of ectoplasm "materializing" from a medium, and it's clear it was bunk. And not even well-executed bunk. As ghost-biter Gallagher so eloquently put it: "That there stuff is just gauze!" What was going on? What happened to the minds of science that they would, even for a moment, buy into this?

The admissions office lady calls me in and asks for my ID

and an "academical letter of introduction." I hand her a print-out of an e-mail from the Keeper of the SPR Archives (*Dear Madam, . . . The Alleged Ectoplasm is not a pleasant object, I should warn you!*). And that's it. I'm in.

The reading room is on the third floor. It has ceilings all the way up in heaven and enormous multipane windows with benedictive shafts of light angling down onto the students. The woman across the table is hunched over a notebook, translating and transcribing from an ancient bundle of brittle blue airmail letters penned in Hebrew. The youth to my left is sacrificing his vision and social life to medieval land transfers. A page arrives with my requested materials: six files, a photo album, and a box containing the Alleged Ectoplasm. The box is made of decoratively patterned cardboard and tied up with a piece of string, like something brought home from a pastry shop. It is larger and showier than I expected. I set it on the floor before anyone can ask what's inside. The plan is to open it later, when my tablemates have left for lunch.

The top file is labeled "Goligher, Kathleen." This file goes with the photo album, which contains photographs of Miss Goligher and her visible ectoplasms, dated 1920–1921. Goligher was the inspiration for the theories and experiments of Dr. W. J. Crawford, a lecturer in mechanical engineering at Queen's University of Belfast. The Golighers were a family of spiritualists—all four daughters worked as mediums—known for their lively séances, held in a tattily wallpapered Belfast sitting room. The sittings unfolded in the typical manner of the spiritualist séance. The medium would disappear behind a curtain or inside a medium cabinet. The lights would be put out—mediums claimed light compromised their abilities and damaged the ectoplasm—and a round of hymns and prayers begun. The medium would fall into a trance and begin producing "demonstrations," as they are still called today by spiritualists, of the presence and power of whatever spirits he or she had drawn to the room. Most commonly, the spirits would show their stuff by making a table at the center of the séance circle tilt or rise. Since everyone was holding hands, the table appeared to be levitating* without the help of anyone seated in the circle.

While most spiritualists were satisfied with the explanation that it was the energy of their spirit friends that was causing the table to rock and rise, Crawford wanted to know exactly how, and by what scientific principles, the dead were accomplishing this. Following a series of experiments involving scales and pressure sensors, he came up with a theory of bendable ectoplasmic rods and cantilevers, which he set forth in embarrassing detail in the 1920 E. P. Dutton hardcovers *The Psychic Structures at the Goligher Circle* and *Experiments in Psychical Science*. (Such was the gullibility of the day that Dutton published five of Crawford's books without a modicum of queasiness. A 1919 letter in the Dutton archives describes Crawford's discoveries as being "of such enormous importance to physics that it is not too much to suppose that they may shew the way to a complete upsetting of present ideas and the building of a new theory on the constitution of matter."†)

Applying laws of physics and engineering, Crawford cal-

---

*Often the medium was using her foot to manipulate the furniture. However, Spirit Table Lifting aids were available for $12 by mail-order through the likes of the Ralph E. Sylvestre Company ("our effects are being used by nearly all prominent mediums," brags the 1901 catalogue). Other helpful items included Telescopic Reaching Rods, self-playing trumpets, and Luminous Materialistic Ghosts ("appears gradually, floats about room and disappears").

†They did not, however, shew the way to a new theory on royalty rates. Dutton courteously rebuffed Crawford's request to double his royalties to 20 percent.

culated that for tables up to thirty pounds, an unsupported or "true" cantilever was employed. The ectoplasmic rod issues "from the torso" (more about this euphemism to come) of the medium, travels parallel to the floor until it reaches the center point of the table, and then changes direction, rising up in a column of four-inch diameter to meet the underside of the tabletop, where it grips the surface by means of a suction-based grasping organ. (He claimed to be able to hear the "suckers"—never a good word choice for W. J. Crawford—"gripping and sliding over the wood.") If the table weighed more than thirty pounds, Crawford concluded, a supported cantilever was employed by the spirits: It angled down and met the floor, which it used for support, before bending and rising to meet the table.

Crawford attempted to prove the existence of the two types of ectoplasmic cantilevers by setting a pressure-recording apparatus under the séance table. This consisted of two pieces of wood that, when pressed together, would complete a circuit, causing a bell to ring. Crawford describes telling Goligher's spirit helpers—or "operators"—to begin a series of levitations, first with a true ectoplasmic cantilever and then a supported ectoplasmic cantilever. As he predicted, the bell rang only when the supported cantilever was in use.

Crawford didn't stop there. He wanted more proof. He wanted the sort of proof he could carry out of the séance room and show to his colleagues. Which brings us to Experiment 4—"Impression on modeler's clay of bottom of cantilever column"—and to the beginning of Crawford's tragic demise. Here is Crawford describing the experiment:

> I brought a little box filled with soft clay to the séance room, and said to the operators, "You remember some time ago when we were investigating the meth-

> ods by which you levitate the table, I found that if necessary you could levitate it by putting the bottom end of the columnar part of the cantilever on the floor immediately under the table, so that it forms a kind of prop?" . . . Answer—"Yes." "Well, I am going to place this box of soft clay under the table and I want you to levitate the table by this method." . . . In a very short time, the table levitated immediately above the clay. . . . At its conclusion I examined the clay. There was a large irregularly shaped impression on it, the length one way being about 3 inches and the other 2½ inches.

In the Goligher file on the table in front of me is a mounted black and white photograph of a similarly obtained "cantilever" impression. It was not made by Crawford, but by a contemporary of his, a psychic researcher named Henry Bremset, who attempted to replicate Crawford's experiment at a different séance circle. He filled two shortbread tins with putty, placed them under the séance table, and explained to the medium what he hoped to achieve.

Tin Number 1 revealed, said Bremset in a letter to the SPR, "the perfect imprint of a woman's shoe." Tin Number 2 bore the imprint of a stockinged heel, "showing stitch marks of knitting." The second photo in particular made Bremset feel a little ill. For he recalled that Crawford, in one of his papers, had described an impression that suggested that of "a gigantic thumb with skin, and lines similar to a human thumb [print]." Bremset sent his Tin Number 2 photo to Dr. Crawford, to see how he'd interpret it. The possibility that Bremset's imprint had been made by a foot in a stocking appeared not to have entered Crawford's mind. His reply stated that he hoped to prove that the impressions were made by psychic rods that carried the impressions of the area from which they emerged.

Perplexed and concerned, Bremset took a train to Belfast later that week. He describes his visit with Crawford in the SPR letter. "I had a long and earnest discussion with him about the interpretation of the facts as I saw them. He was obviously profoundly disturbed though still clinging to his new theory. . . . When I parted from him he looked a very worried man. . . . Not long afterward came his tragic death."

Crawford drowned himself in the summer of 1920. Though his suicide note stated that his actions had nothing to do with "the work," hearsay held that the motivating force for the suicide was a profound embarrassment upon realizing he'd been duped.

The real mystery, as far as I'm concerned, is that it took him so long to figure things out. In his book, he has included a photograph captioned "The cantilever method of levitation. A rough cantilever in position." Yet it unequivocally depicts a strip of filmy white cloth—coming down from Goligher's lap, dropping a foot or so, and then curving up to the bottom of a small desk. I don't know anything about engineering, but it's clear to me that that material isn't "supporting" anything. It's just hanging there limply. There's nothing mysterious or suggestive about it. Crawford's album contains photo after photo of Goligher with lengths of "the substance" lying on her lap, wadded on the floor at her feet, tied to the table legs, and in every case it's impossible to mistake it for anything other than ordinary man-made cloth. One photo shows Kathleen Goligher with a pleated bunch of fabric near the neck of her white shirt, and I spent a good minute and a half trying to decide if it was fashion or ectoplasm. Many of the album photos are reproduced in Crawford's books, but not photos 4F or 5E: Here is Goligher, her standard eyes-closed, medium-in-trance pose abandoned for the moment, openly laughing or grinning.

The other possibility is that W. J. Crawford was—to use the word choice of Harry Houdini, who saw the Goligher photographs and heard the engineer explain them over the course of a three-hour dinner—insane. Evidence for his rather weak grip on reality can be found among the captions in the SPR album I've been looking through today. Photograph 8E, for instance: "In this photograph are to be seen the white and grey substances. Dr. Crawford said that the grey substance left excreta marks." On June 22, 1920, shortly before he died, Crawford wrote in his journal that he was considering the possibility that ectoplasm emerged from the medium's rectum. He first arrived at this unorthodox notion upon finding "particles of excreta" in the white drawers that he asked the medium to put on before—and return after—the séance. It takes a certain kind of mind to interpret smidgens of fecal matter found in underwear as an ectoplasmic calling card rather than an ordinary by-product of a minor lapse in hygiene. It takes, I would think, a mildly psychotic kind of mind. Crawford's distinctive psychosis appeared to include a troublesome underwear fixation. In addition to the white drawers, we find the following "highly probable facts, resting on good authority" in a letter from a Mr. Besterman, in the SPR archives. Shortly before Crawford's suicide, Besterman writes, Crawford "spent all his money (consequently leaving nothing) on a stack of woollen underwear for his family, sufficient to last for several years."

After Crawford's death, the SPR sent another researcher, E. E. Fournier d'Albe, to follow up with the Golighers. Though Fournier d'Albe had earlier in his career vouched for the authenticity of the original ectoplasm-exuding medium Eva C., he was suspicious of Goligher from the start, largely because of Crawford's photos. In his ninth sitting at the Goligher circle, on June 23, 1921, he confronts the spirits about the ectoplasmic cantilevers in Crawford's photographs: "Well, I cannot make out this structure. In some places it appears as if woven. Have you a loom in your world?"

Shortly thereafter, Fournier d'Albe caught Kathleen Goligher levitating a stool with her foot. Convinced that her ectoplasm could be bought by the yard in downtown Belfast, Fournier d'Albe purchased a yard of fine chiffon, and published a shadowgraph close-up of it, run side by side with a shadowgraph of a bit of "ectoplasmic rod." The two appear identical.

Along with putting the Golighers through their paces, Fournier d'Albe read through Crawford's correspondences and unpublished séance reports. Time after time, Crawford misinterpreted straightforward evidence that Goligher's "psychic structure" was her right foot. "Touching end of structure," read Crawford's notes from a séance in October 1919. "On one occasion the part I felt was *like bones, close together, like finger bones bent over . . . or toes of feet and even the nails.*" If Crawford was at all suspicious, he made no mention of it.

Citing nerves, Kathleen Goligher retired from mediumship in 1922, upon the publishing of Fournier d'Albe's book. The SPR file includes an envelope of snapshots from a rare Goligher sitting given fifteen years later, the result of tireless cajoling on the part of a researcher named Stephenson. Positioned in front of Goligher is a crudely constructed wood and chicken-wire cage, which Stephenson appears to be using to trap the ectoplasm. Kathleen looks older than her years. Her head is bowed, and her hands are clasped in her lap. No one is smiling. If not for the rabbit hutch parked at their feet, they could be the bored guests of an especially tiresome tea. Even the ectoplasm, unfurled limply on the carpet like painters' rags, looks weary of it all.

While Kathleen Golighe relied on Crawford's credulity to make a name for herself and her fabric-store emanations, Boston medium Margery Crandon managed to fool the best and the brightest. In 1924, *Scientific American* offered a $5,000 prize to any medium who could produce a verifiable "visual psychic manifestation." The medium would have to demonstrate her talents before a committee of investigators chaired by *Scientific American* staffer Malcolm Bird and consisting of Harvard psychologist William McDougall, Massachusetts Institute of Technology emeritus Dr. Daniel Comstock, Society of Psychical Research officers Walter Prince and Hereward Carrington, and prestidigitator and tireless medium debunker Harry Houdini. The only medium found worthy to sit before the committee was Boston's Margery Crandon, the wife of a Harvard-educated obstetrical surgeon and the cause of great, protracted ballyhoo over at the American Society for Psychical Research. (The ASPR, now in New York, started out in Boston.)

Twenty séances later, the *Scientific American* committee was hotly divided in its conclusions. Houdini and McDougall believed her to be a fraud. Comstock and Prince waffled, saying that although Margery had failed to prove herself, more data were needed. On the other side of the fence, Bird and Carrington declared their belief that her phenomena were genuine. (Both Bird and Carrington were accused of turning a blind eye—or even being party to the deceit—for reasons of personal financial gain, in the form of book royalties and lecture fees.) McDougall and Houdini pointed out that the more thoroughly constrained were Margery's hands and feet, the less likely she was to produce ectoplasm. "The more care, the less wonder," as McDougall put it. Houdini at one point built a special cabinet-box for her, similar in appearance to those 1960s steam cabinets in which villains would lock James Bond and spin the temperature dial to max. In the end, the committee voted not to award her the $5,000. Bird was eventually called to task by *Scientific American* editorial chief O. D. Munn, who pulled the latest Bird piece from the magazine at the last minute. I haven't followed the course of *Scientific American*, but Bird's earlier straight-faced seven-thousand-word blow-by-blow of a Margery séance would seem to be a low point.

The Margery ectoplasms were of an entirely different species from those of Kathleen Goligher. "The appearance is somewhat that of a sheep's omentum," reads the caption of a photo in the ASPR files. (An omentum is a curtain of fat that hangs down from the stomach and insulates the intestines. In actual fact, the material came from a sheep's lung—or so concluded a team of Harvard zoologists and biologists to whom McDougall submitted the photograph for analysis three years later.) The photograph shows a pair of studious-looking men in bow ties and spectacles leaning in close over a séance table to scrutinize a singularly unappetizing mound of alleged ectoplasmic matter. Margery's torso appears in the background, clad, somewhat incongruously, in a satin floral print dress stretched tight over her own, rather well-developed omentum. Plate 2 from the same set shows the medium slumped forward onto the séance table, looking as though she'd been shot in the head, the "matter" now poised upon her neck and ear. In Plate 14, the ectoplasm is shown escaping from Margery's nose, whereupon it was said by the medium to assume the form of a "tracheal speaking appendage," used by Walter—Margery's dead brother and now spirit guide.

Though the Margery ectoplasms seemed content to enter the world through any handy orifice, most often they emerged from between her legs. As in Plate 5: a "crude teleplasmic hand, originating from the genitals." Based on a casual survey of the literature on all the materializing mediums, the vaginal canal was the most common ectoplasmic exit strategy. Indeed, some months before Crawford embraced his rectal theory, he posited that the substance might be issuing "from inside the legs." And so Crawford, in inimitable Crawford style, devised an experiment involving special underpants. "The medium put on white calico knickers under my wife's supervision," he wrote in *The Psychic Structures at the Goligher Circle*. "Carmine powder was placed in her shoes. At the end of the séance it was found that there were carmine paths up to the top of both stockings and then *inside* the legs of the knickers to the join of the legs. . . . Thus, as I had expected for some considerable time, it was abundantly clear that the plasm issued from and returned to the body of the medium by way of the trunk." Why the ectoplasm would have felt the need to visit the inside of the medium's shoes before its return trip "between the legs" is a mystery Crawford did not address.

And now I'm going to pass the microphone to William McDougall. For how many chances do we have to hear a Harvard professor hold forth on vaginally extruded ectoplasm? "There is good evidence that 'ectoplasm' issues, or did issue on some and probably all occasions [from] one particular 'opening in the anatomy' (i.e. the vagina)," allowed McDougall in his summary statement for *Scientific American*. "The more interesting question is—How did it come to be within 'the anatomy'? There was nothing to show that its position there and its extrusion from that place were achieved by other than normal means." In other words—please forgive me—she stuck it up there, and then she pulled it out.

The debate over Margery and her ectoplasms raged on for a full year. Some wondered how she could possibly have room in her womanly interior for the array of objects often produced during séances. And it was at times an impressive array: In a 1925 letter from conjurer Grant Code, the medium is described as having been caught "drawing from the region of the vulva two or three objects which were exhibited on the table as Walter's hands and terminals." Code himself found it difficult to imagine how she managed it, and wondered whether Margery's husband—who was after all an obstetric surgeon, a veteran of some one-hundred-plus cesareans—might have carried out a surgical enlargement of, as he put it, "Margery's most convenient storage warehouse."

With that, the debate deteriorated into name-calling and threats. Crandon counters Code's implications with accusations that Code had raped his wife at a séance. The SPR's Dr. Prince, in defense of Code, writes that Dr. Crandon was dismissed from his most recent position over the "systematic seduction of nurses." Margery threatens Houdini with "a good beating." Even the discarnate Walter joins the fray, calling Dr. Code "a boob." The most damning letter of all comes from McDougall's colleague J. B. Rhine, who was soon to put paranormal research on the more strictly experimental—if vastly less entertaining—track to card-guessing and dice-tossing. (Rhine founded Duke University's famous Parapsychology Laboratory.) Here is J. B., sounding the much-needed voice of reason:

> We left the house feeling we had witnessed nothing but a daring though artfully concealed attempt to capture notoriety. Why must we sit in darkness, while Dr. Crandon may, unannounced, flash on his white flashlight . . . ? Why, if light injures the structures, should [the alleged spirit entity] Walter seize the luminous end of the megaphone, placing his "grasping organ" right over the luminous band? Why is it that for certain acts, Dr. Crandon must be next to the medium "for her protection"? Why do they refuse to allow one

> to place one's fingers lightly on the medium's lips to test the independence of Walter's voice? . . .

Returning to the matter of the warehoused ectoplasm. As regards the feasibility of such a practice, it is worth pointing out that Margery wouldn't be history's first vaginal smuggler of bulky carcass parts. In 1726, a rumor spread through England about a peasant woman from the outskirts of Guildford, who was giving birth to rabbits. (The story is spun out in precise and rollicking detail by medical historian Jan Bondeson, from whose remarkable book *A Cabinet of Medical Curiosities* come these facts.) The rumor soon made its way to the Prince of Wales, who, fascinated,* promptly dispatched the court anatomist, Nathaniel St. André, to investigate. St. André, an ambitious self-promoter with no real medical training, arrived to find Mary in labor, about to give birth to her fifteenth rabbit. The fourteen siblings, all stillborn, were on display in jars of alcohol, arranged by Mary's proud man-midwife, John Howard.† Minutes after the bewigged St. André entered the room, the forward half of a skinned four-month-old rabbit dropped into Howard's receiving blanket. Howard conjectured to St. André that the rabbits were being crushed into pieces and skinned by the force of Mary's contractions. Later that night, Mary "gave birth" to the back half of the animal—Bondeson

---

*Gynecological preoccupations are a running theme with the Princedom of Wales. Two and a half centuries later, the Prince of Wales would be caught in an intercepted cell phone call voicing his desire to be reincarnated as his lover's tampon.

†The man-midwife, with his arsenal of forceps and knives, was a recent arrival on the obstetrical scene and much resented by the gentle guild of midwifery. "Yea, infants have been born alive, with their brains working out of their heads, occasioned by the too common use of instruments," warned midwife activist Sarah Stone in her 1737 *A Complete Practice of Midwifery.*

describes Howard and St. André studiously putting the halves together and deeming it a perfect fit—and, later still, its skin.

A postmortem, performed by St. André's staff back at the court, uncovered pellets of "common rabbit Dung" in the rectum, an obvious indication of fraud that went unnoted by St. André. The ignorant anatomist vouched for Mary's authenticity, and the prince ordered the peasant woman brought to London, where she and Howard enjoyed a brief spell of fame and (relative) wealth. Unfortunately for Mary, one of her London visitors was the respected obstetrician Sir Richard Manningham. When Mary tried to pass off half a hog's bladder as her placenta, Manningham—you have to love this guy—came back the following day toting a fresh hog bladder for comparison. Whereupon Mary, having no good explanation for why her placenta carried the "strong urinous Smell peculiar to a hog's bladder," burst into tears.

Mary Toft's final downfall came at the hands of a porter at her lodgings. Unable to procure rabbits in central London, she had tried to bribe a porter into tracking some down. The porter talked, and Mary eventually confessed. She explained that when the doctors' backs were turned, she would transfer into her birth canal a rabbit, or rabbit portion, which she had had concealed in a special "hare pocket" inside her skirt. Whether John Howard was in on the hoax or simply another victim of it was never clear. What is clear is that male medical professionals could be ruinously susceptible to vaginal deceits.

IT'S 1:40 P.M. now, but no one at my table has left for lunch. I pick up the box of ectoplasm and rest it on my lap. It's worse than I thought. Slipped under the string is a three-by-five card, upon which is typed the official archive summary:

> Material alleged to have been captured from Mrs. Helen Duncan, materialising medium, at a seance in 1939. . . . She had been stripped and searched but with no vaginal examination. The material was smelling and had bloodstains on it which appeared at regular intervals. The suggestion was that the blood had soaked into the material while it was folded up, and that the most likely explanation was that it had been secreted in the vagina.

Inside the box, a yellowed paper envelope is tied with a length of pink bias tape. It's a large envelope, bigger and heavier than a four-month-old rabbit. I would put the weight at close to a pound. That is a lot of stinky ectoplasm. It's a lot of stinky ectoplasm to spread out and examine in the still, reverent hush of the Cambridge Manuscripts Reading Room. I want to smuggle it out of here and open it up in the ladies' room, but my bag is checked in a locker downstairs, as per manuscripts room rules. Oh, for a hare pocket.

I turn to the Helen Duncan file, in the hopes that by the time I am done reading, the people at my table will have fainted from hunger or gone home. Duncan was ectoplasm's last stand. And what a stand it was. A histrionic Scotswoman of poor health and bad habits, Duncan weighed close to 250 pounds. She smoked constantly and moved with obvious difficulty, often requiring assistance to rise from her seat and make her way across the séance room. She had nine children, who hung from her hems and scaled her bulk like small mountaineers. One biographer described the youngest child atop her lap, dandling the flesh that hung down from her massive upper arms. Her séances were high drama. She tended to swoon and fall off her chair and occasionally wet herself in the frenzy of spiritual possession. She once emerged from the séance cabinet naked under a floor-length "ectoplasmic veil." For those whose interest in spiritualism was purely voyeuristic, Helen Duncan was the hottest ticket in town.

Duncan produced ectoplasm as readily and lustfully as she produced offspring. However the two did not typically—item SPR 197.1.6 notwithstanding—issue from the same anatomical opening. Owing to the well-publicized stunts of Margery and other 1920s mediums, those active in the 1930s were subjected to thorough body cavity searches by researchers before each séance. "Thorough" meaning:

> May 14, 1931
> After the séance room and cabinet had been examined, the medium was led into the room by Mrs. A. Peel Goldney. . . . The doors having been locked, the medium was placed upon a large settee . . . and in the presence of Dr. William Brown, Mrs. Goldney (who has trained and worked for many months in a midwifery hospital) made a thorough vaginal and rectal examination. The rectum was examined for some distance up the alimentary canal and a very thorough vaginal examination given.

This passage, written by magician-turned-psychic-researcher Harry Price, describes preparations for a séance undertaken in Price's National Laboratory of Psychical Research (NLPR) in London, part of a two-month investigation of the Duncan mediumship. Price covered all the angles. He designed a special fraud-preventing "séance garment" that enrobed the entire medium, including her hands and feet, such that only her head stuck out. So even if Mrs. A. Peel Goldney had managed to miss something in her anatomical inspections, it would have been impossible for Helen to get that something out of the suit and into the open. Price's book about the Duncan investigation includes a dozen or more photographs of the medium ensconced in her special garment. It is fashioned from satin in a loose jumpsuit style, which, in combination with Mrs. Duncan's sizable mid-torso circumference, brings to mind late-career Elvis, or the sad clown in that Italian opera. I should point out that Mrs. Duncan was compensated for her humiliations at the NLPR. Handsomely so—five hundred pounds in all. This helps explain the medium's seemingly inexplicable decision to risk her career in the laboratories of the NLPR.

Price was surprised and confounded to see that Helen Duncan was able, despite his precautions and within minutes of the séance beginning, to produce a six-foot-long ectoplasm. "The séance garment should absolutely preclude the secretion in or extraction from the orifices I have mentioned, even had she not been examined medically." Forced to rule out "the vaginal-cum-rectal theory," he came up with an equally extraordinary possibility: "That the medium possesses a false or secondary stomach (an esophageal diverticulum) like the rumen or first stomach of a ruminant, and that she is able to swallow sheets of some material and regurgitate it at leisure—like a cow with her cud."

This was not as far-fetched an idea as it sounds—particularly in Price's day. Search the British medical journals from the early 1900s, and you will come across lengthy articles on the subject of human ruminants: seemingly ordinary citizens who could effortlessly "bring up" portions of their most recent meal for further mastication and—quite often—enjoyment. "It is sweeter than honey, and accompanied by a more delightful relish," a Swedish ruminator is quoted as saying in

E. M. Brockbank's "Merycism or Rumination in Man," which ran in the February 23, 1907, issue of the *British Medical Journal.*

No one could say whether the condition was hereditary or learned. Brockbank cites the case of a tin worker as support for heredity's role. "He looked upon it as a perfectly natural phenomenon, descending from his grandfather and father to himself, and to all of his sisters and brothers and to many of their children. . . . [His wife], a bright intelligent woman who does not ruminate, states very definitely that as soon as the children began to walk they used to bring up mouthfuls of food, which at first they spat out, later they began to rechew it, especially after a meal they liked." Other physicians insisted the habit was passed along by imitation, citing as evidence a Swiss ruminator who lived among cows all his life, and a boy who was suckled for two years by a goat, and "acquired by imitation his foster-mother's . . . habits."

Though the act appears identical in cow and man, only in the bovine does it serve any useful purpose. Though occasional exceptions did exist, such as this 1839 *Lancet* case study of a farmer: "To save time, he had acquired a habit of 'bolting' his food . . . then getting on horseback, and subjecting his dinner piecemeal to mastication at his leisure." The farmer didn't seek medical advice until later in life, after falling into some wealth and attempting to mix with a higher cut of society, who found his habit "very disgusting." Two papers I read implied that ruminating was accepted as normal behavior among the working class, implying that cud chewing was as common among nineteenth-century laborers as tobacco chewing among modern-day major league pitchers. These days, rumination articles are confined to literature about psychologically or developmentally impaired individuals.

(Happily, there is help. A surgical technique recently perfected at the Swallowing Center at the University of Washington* stops rumination in its tracks.)

Nor is it true that, as Harry Price suggested, human ruminants possess bovine-style multiple stomachs. This was a stubborn rumor fueled by two seventeenth-century cases of ruminating men with horns—one a unicorned Paduan nobleman and the other a bicorned monk. Autopsies of ruminants—whose stomachs were normal—put a stop to the rumor, as did a paper by a physician named Sachs, who reviewed one hundred cases of men with rudimentary horns and found only one ruminant among them.†

So Harry Price was wrong to surmise that Helen Duncan was ruminating ectoplasm that she stored in an auxiliary stomach. Duncan's was more likely a case of masterful regurgitation. Regurgitation acts were a sideshow mainstay in Price's and Duncan's day. In his book *Regurgitation and the Duncan Mediumship*, Price describes regurgitators of live goldfish and snakes, light bulbs, razors, pocket watches, bayonets, two eighteen-pound dumbbells, and a rolled umbrella. Colleague Harry Houdini watched a frog-swallower in Warsaw swallow thirty or forty glasses of beer

---

*As opposed to the Swallowing Center at Northwestern, or the Swallowing Center at the University of Southern California, or the one at Holy Cross, or the Rusk Institute, or the Nebraska Medical Center. Of course, the original "swallowing center" is a chunk of your brainstem that coordinates chewing, gagging, vomiting, coughing, belching, and licking, all with minimal fuss and no funding from the NIH.

†I once saw a wax model of a horned human head at the Mütter Museum in Philadelphia, but I had no idea the condition was sufficiently common for a doctor to pull together one hundred cases for a review paper. But what do I know? Perhaps horns were the plantar warts of their day. Perhaps Sachs held a post at the Horn Center at the University of Padua.

and an unspecified number of half-grown frogs, which he would then bring up alive. I'm unclear on whether the beer helped with the process or with the man's state of mind, or possibly that of the frogs.

Thus it is within the realm of possibility that Helen Duncan was swallowing and regurgitating sizable rolls of cheesecloth. To demonstrate the convenient compactibility of this fabric, Price bought a six-foot by thirty-inch swath, rolled it up tight, and photographed his secretary Ethel with the fabric sticking from her mouth like a Mafia gag.

Far more damning than the Ethel photo was Mrs. Duncan's tantrum in response to a request, on May 28, 1931, that she submit to a post-séance X-ray. Price wanted to find out if she had an extra "pseudo-stomach," and/or what was in her stomach(s). He was aware that the chances of getting a clear image through "the depths of the medium" were slim (early X-ray technology being what it was); but she was not. As the equipment was readied, Mrs. Duncan suddenly leapt from the settee, bowled over Dr. Brown, pushed aside Mrs. A. Peel Goldney, and lumbered screaming into the street. Her husband (and long-suspected accomplice) ran after her, and the two were gone for ten minutes, during which time—Price and his team suspected—she regurgitated the fabric and passed it off to him. And what a sight that must have been for genteel passersby—a panting, hysterical woman in a clown suit, throwing up a roll of cheesecloth.

Upon their return to the lab, Helen, visibly calmed, agreed to—nay, insisted upon—the X-ray. Price, no fool, took Mr. Duncan aside and asked him if he would object to being searched. Mr. Duncan did object, "murmuring something about his underclothing." It always comes down to underpants with these guys.

More support for the regurgitation theory comes from the research department of the London Spiritualist Alliance, which conducted some of its own investigatory séances with Mrs. Duncan. On June 12 of that same summer, Helen was asked to swallow a pill containing methylene blue, so that anything regurgitated would be marked by the dye. No cheesecloth appeared that night (though the medium at one point attempted to pass off her tongue as ectoplasm).

Two weeks after the last Duncan séance, the council of the NLPR called Mr. Duncan in to a meeting and confided their suspicions. They were well armed. Price had with him a detailed and damning eleven-page lab report of a chemical analysis of a cutting of Duncan ectoplasm, which Mrs. Duncan's spirit guide Albert agreed to make available. (Price describes this séance as resembling a sewing bee, with its seated circle of men and women, all poised with scissors, awaiting Albert's go-ahead.) The council then showed Duncan a photograph of his wife draped in her ectoplasm side by side with a photo of Price's ever-game secretary Ethel similarly posed and draped with a length of Woolworth's cheesecloth. Duncan was unable to tell the difference. Finally, at a "heart-to-heart" on June 22, Mr. Duncan concurred that it was likely that the ectoplasm was produced by regurgitation, though he insisted that it was "subconscious regurgitation."

"We pointed out," writes Price, "that . . . in that case she would have to . . . buy the cheese cloth subconsciously . . . and swallow the bag subconsciously." The aptly named Dr. Price offered Mr. Duncan one hundred pounds to convince his wife to be filmed in the act. Duncan promised to do what he could, but the couple lit out for Scotland the following morning.

The ectoplasm in the box at my feet is dated 1939, so Helen got up to her tricks at least once more. It's possible the sample is one of the last of its species. It had been three years since Margery had been coaxed out of retirement for the dispirited rabbit-hutch sitting. Kathleen Goligher had long since disappeared from the scene. There hadn't been an article about ectoplasm in the *New York Times* for twelve years. For all I know, this is the last sample ever produced, ectoplasm's Ishi.

Inside the box is an envelope tied with a length of pink bias tape. I take it out and place it on the table. I pull one end of the pink ribbon, slowly and with drama, like a man uncorseting his lover. Rather than the more typical and compactible cheesecloth or chiffon, it is some kind of cotton with a satiny finish. The stains are faint and brown, the smell manageable but detectable. I unfold it to get a rough idea of size—I'd say ten feet by three feet. It's huge. It's as though the Keeper of Manuscripts and Archives came in drunk one day and got the Shroud of Turin mixed up with Helen Duncan's ectoplasm. The Hebrew reader glances up, then returns to her work without comment.

I don't care how many children marched down the Duncan birth canal; I have a hard time believing any woman could "secret" this much fabric through that size opening. Barring a visit to the surgical practice of Boston gynecologist Dr. Crandon, my guess is that Mr. Duncan—who insisted on being seated next to his wife at séances—slipped it to her undetected.

Despite his travails with the Duncans, Harry Price did not give up hope that some mediums were possessed of genuine powers. Price's book concludes with an optimistic pronouncement about the authenticity of the medium Rudi Schneider, known for ejaculating during particularly heady séances. (To my surprise and his credit, Schneider did not try to pass off the ejaculate as ectoplasm.) I don't have the full story on Schneider, nor am I going to go dig it up, because I want to get back to the present. Fast-forward to the NLPR of 2004: The

University of Arizona Human Energy Systems Laboratory, where they test modern-day mediums.

I put the ectoplasm back in its envelope, tie the pretty pink ribbon, and return the box to its keeper. By 8 p.m., I'm back in London, at a Pakistani restaurant down the street from my hotel. In honor of Margery Crandon, I order lamb.

ABC
NO

6

The Large Claims of the Medium

Reaching out to the dead in a University of Arizona lab

GARY SCHWARTZ is an uncommon hash of academe and spirituality. He has a Phi Beta Kappa key from Cornell and a previous tenure at Yale, but he is best known for his laboratory tests of mediums (the subject of his book *The Afterlife Experiments*). He is a psychology professor at the University of Arizona,* as well as the founder

*Where, says the school's vice president for research, "professors are allowed to pursue their own interests as long as there's nothing unethical or illegal." It's a sentiment echoed by the powers-that-be at no less than Harvard Medical School, which counted as faculty the late alien-abduction researcher/sympathizer John Mack. "Oh, they're all weird or embarrassing one way or another," said a spokesman some years ago when I expressed surprise at Mack's tenure. "Besides," he added ominously, "there's a lot of weird science that turns out to be not-so-weird once it's proven."

Schwartz's $1.8 million NIH grant, of which the university takes fifty-one percent, helps ease any discomfort over the lab's unorthodox research.

of the university's Human Energy Systems Laboratory. Waiting for him in my hotel lobby, I did not know whether to be looking for a man in a tweed jacket or a man in drawstring trousers or—help me—a man in both. Schwartz further muddied the waters by showing up in a double-breasted suit, with a white Jaguar parked in the lot. Whatever the heck he's up to, he's doing well with it.

Schwartz's experimental data have led him to the conclusion that there are people—rare and gifted mediums—who can communicate with people who have died. Unlike researchers I have met at the University of Virginia, the other American university currently hosting research on the paranormal, Schwartz will comfortably and without reservation voice his conclusions regarding the hereafter. He tells me, as we drive along Tucson's glarey, succulent-accessorized streets, that he is working on a paper for *American Psychologist* called "Reexamining the Death Hypothesis." As in, no one has actually *proved* that the death of a human body is the end of the trail for the personality that lived in it. It's just a hypothesis.

Like the medium researchers of the early 1900s, Schwartz brings to his laboratory only the most talked-about mediums. John Edwards (of *Crossing Over*), for example, was one of his "dream team" of mediums tested in the *Afterlife Experiments* work. His most recent discovery is Allison DuBois (the medium upon whose life NBC's *Medium* is based).

Perhaps because I'd been reading a biography of the slovenly and bellicose Helen Duncan on the plane to Tucson, it did not cross my mind that a medium could look like a beauty pageant winner. DuBois has long, obedient rust-red hair that turns up just so on the ends and complements her coppery lipstick. Her blush and foundation could have been applied by airbrush, so perfect is the blending. She manages to look made-up at the same time as she looks completely natural and beautiful without device. I can no more understand how a woman does this than I can understand how a woman communicates with dead people. DuBois is paranormally good-looking.

Before she became a medium, Allison DuBois was on a career track to the Maricopa County prosecutor's office as a criminal prosecutor. In April 2000, someone got in her way. Well, more than that. "I went downstairs to get the laundry, and a man walked through me," she told me at lunch. She was mounting an enthusiastic attack on an apple tart as large as her foot, which isn't large as feet go, but very large as apple tarts go. DuBois's delicate frame belies a robust enthusiasm for food, which seems to run over into her readings, which often include food preferences of the "discarnate," as they say in these parts. The discarnate man in the laundry room being no exception: "I knew that he loved clam chowder, and that he'd had a heart attack." She ran upstairs and told her husband Joe, who is an aerospace engineer. Joe blinked at her, as one might when abruptly confronted by one's spouse's possible mental disintegration, and then he said, "That's my grandfather."

After a few months of running into dead people, DuBois saw a *Dateline* segment about Gary Schwartz's medium research. "I thought, 'I'm going to go see him. I'm going to prove to myself that I'm not really doing this, and then I'm going to get on with my career.'" DuBois impressed Schwartz as having genuine talent. The prosecutor's office would have to wait.

DuBois is one of four mediums taking part in a research project called the Asking Questions Study. I love this study, because it addresses one of my main beefs with medium-brokered encounters with the dead. Dead people never seem to address the obvious—the things you'd think they'd be bursting to talk about, and the things all of us not-yet-dead are madly curious about. Such as: Hey, where are you now? What do you do all day? What's it feel like being dead? Can you see me? Even when I'm on the toilet? Would you cut that out? If the dead come through at all, they come through in cryptic little impressions: a stout woman with gray hair, a small black dog, the date May 23. It's a maddening way to communicate. Schwartz and his mediums would reply that that's the best the dead can manage, that they can't speak sentences into the medium's head. Impressions come through, and that's all.

Julie Beischel, the researcher behind the Asking Questions Study, wondered whether perhaps it was the case that the dead never provide this kind of information because no one ever bothers to ask them. Beischel is a University of Arizona psychology postdoctorate who, like DuBois, contacted Schwartz after seeing him on TV. Beischel was a pharmacology student whose interest in the paranormal was sparked after losing her mother. (This is a common theme among people involved in paranormal research. As she puts it, "Everyone doing this has lost somebody." Certainly this was the case with some of spiritualism's less likely converts: Sir Arthur Conan Doyle and physicist Sir Oliver Lodge both lost sons during World War I.)

Beischel assembled a list of thirty-two questions about the afterlife, which are being posed to two discarnates, via four mediums. (Each medium takes a turn with each of the two dead folks.) Beischel hadn't analyzed the data at this point, but she gave me printouts of the answers she had collected. With both discarnates, the answers to a given question usually differed with each different medium. "Do you eat?" for example, garnered an even split of yes's and no's. I asked Beischel how she interpreted this. She said, very straightforwardly, as is her manner: "My interpretation is that the mediums are just guessing, or the answer is biased by the medium's own ideas of what the afterlife is like, or the questions don't have enough emotional interest for the discarnate to give a strong answer." Which more or less covers all the bases. In case the answers are in fact coming from the Beyond, I've culled some highlights for you, from various mediums. We'll start with the good news:

Q. *What do you do every day?*
A. She's showing herself at the table eating.
Q. *What type of "body" do you have?*
A. She says fat people are thin here. . . .
Q. *How is the weather?*
A. It's Florida without the humidity.

And now, some less good news:

Q. *Is there music?*
A. Yes. She whips out a xylophone and goes, bum, bum, bum bum bum. And I also get the Carpenters.
Q. *Are there angels?*
A. Yeah . . . but they've got their own clique going. They've got their own little deal going on.
Q. *Do you engage in sexual behavior?*
A. I don't know if, like, she can and she chooses not to or what the deal is, but it's like, no, not really.

And a point of interest for aspiring writers . . .

> He's showing me writing. [Experimenter: He's writing a book?] I don't know. I mean, understanding the fact that there are no, you know, physical constraints, so what the hell, why not, you know? Get your story placed somewhere. I don't know where the hell it would be placed, but somewhere. . . .

To the formal study data, I feel I must add one last statement about the afterlife, passed along to me by Allison DuBois, who received it from an unnamed discarnate during a private sitting: "I can wear pleated pants now."

Though Beischel is the first to have addressed these issues in the purview of a modern university-based research study, afterlife investigators and spiritualists have, on occasion over the years, posed these sorts of questions to the dead. Matla and Zaalberg van Zelst, the Dutch physicists who appeared in Chapter 4 with their soul-proving cylinder-in-a-box gizmo, grilled the occasional *homme-force*, as they termed it, via a medium, about conditions on the other side. The *hommes-force* were, like their interviewers, scientifically minded and tended to provide answers that, while marvelously detailed, did not always get at the things we live humans are itching to know. They'd natter on about the density and specific gravity of an *homme-force*, for instance, or the fact that when they need to move briskly, they assume a spiral form thirty-five centimeters in length and with fourteen turnings, which resembles in shape and hopefully nothing else the type of bacteria found in "*les matières fécales*." You have to wade to the end of the book to find the good stuff, which is nothing short of alarming: Although nearsighted, they can *see through our clothing* and sometimes even through our skin. If the room is quiet, they can read our thoughts.

In the end, I couldn't decide if being an *homme-force* represented an improvement over life on earth. On the one hand, there are no ailments, no need to work or secure lodgings. You own nothing, and you have no cares or obligations. One of Matla's respondents confessed that the life of an *homme-force*, despite the novel joys of eavesdropping on thoughts and peering through ladies' undergarments, soon loses its luster. Because they depend entirely on us for their "distractions," it said, they grow bored during the night while we sleep. And, alas, "*il n'a pas de sexe.*" There is no sex.

This vision of the afterlife as DMV waiting room differs widely from that of spiritualism bigwig Arthur Findlay, who, over the course of three séances with a medium named Sloane, posed fifty-three questions about the afterlife. Various discarnates chimed in with replies. Following is the transcript of a reply to "Will you tell me something about your world?" taken from Findlay's 1955 autobiography *Looking Back.* Basically, everything that exists here on earth exists in the Beyond, except possibly composting:

> We can sit down together and enjoy each other's company. . . . We have books and we can read them. . . . We can have a long walk in the country. . . . We all smell the same aroma of the flowers and the fields as you do. We gather the flowers as you do. . . . Here we have no decay in flower or field as you have. Vegetable life just stops growing and disappears.

Myself, I am hoping that neither Matla nor Findlay is correct, and that the vision of Reverend Dr. G. Owen represents our true future. In a February 5, 1923, *New York Times* article ("Owen Says Heaven Needs Active Men: Sailing One of Pastimes"), Owen laid out the details:

> Big business men are needed in heaven, according to Rev. Dr. G. Owen, an Anglican clergyman who lectured on spirits yesterday afternoon at the Broadhurst Theatre.
>
> The vigilant eye of the expert accountant and the driving energy of the strong executive have ample opportunity for exercise, the speaker said, although

> everything is on an altruistic basis. . . . Doctors are not needed to treat spirit patients, because diseases do not exist, and they are immediately engaged in different specialties and lines of research. . . .
>
> Spirits walk, but some ride about in chariots. . . . "We don't really need them, but sometimes we find them convenient," Dr. Owen said he was told. The clergyman . . . told of placid lakes, rivers and seas and of the inhabitants going about in boats of all kinds, including sailboats. Aerial navigation requires attentive skippership, the clergyman explained, because the spirit body cuts through the attenuated atmosphere at high speed and is steered purely by thought.

The reverend concluded with some "scattered facts about the future," including one that suggests he might have sat down and enjoyed Arthur Findlay's company: "Flowers do not fade. They melt and disappear." The article concludes, somewhat abruptly, with the statement, "We keep our mannerisms."

If that is so, then somewhere, eons from now, Gary Schwartz will be smoothing the tip of his tie over his belly and jingling the change in his pockets. Pacing while he talks. And laughing. Laughing easily, laughing loud, laughing a lot. If the skeptics get under his skin—and I'm sure that they do—he doesn't let it dampen the obvious fun he is having with his work. For a man who takes an extraordinary amount of professional guff, he is resiliently good-natured, a Pooh bear among the skeptic society Eeyores.

Shortly after Schwartz's paper detailing the *Afterlife Experiments* findings ran in the *Journal of the Society for Psychical Research*, University of Oregon Professor Emeritus of Psychology Ray Hyman published a piece in the *Skeptical Inquirer* entitled

"How Not to Test Mediums." The article featured a laundry list of criticisms of Schwartz's methodology. The most damning of these, to my mind anyway, concerned rater bias: Schwartz's sitters had been allowed to rate the readings, and they knew which one had been their own. This is a serious and long-acknowledged bugaboo in research that aims to test mediums and psychics. One of the first researchers to document rater bias among mediums' sitters was a London University Ph.D. candidate named John Hettinger. His project, carried out in the late 1930s, is detailed in parapsychologist Sybo Schouten's excellent "Overview of Quantitatively Evaluated Studies with Mediums and Psychics" in the July 1994 *Journal of the American Society for Psychical Research*. Hettinger found that subjects' judgments of the accuracy of mediums' statements pertaining to their dead loved ones were strongly influenced by their knowing whether or not the statements were intended for them. Cut out the horoscopes in the next astrology column you see, remove the star signs at the top, and mix them up. You'll likely find that yours doesn't stand out the way it usually seems to.

Hyman describes having been a pawn of the phenomenon himself. In his teens, he had earned spending money by reading palms. He started doing it on a lark, and was surprised by his clients' enthusiastic praises of his talents—so much so that he himself began to believe he had a gift. Eventually a friend persuaded him to try an experiment in which he would deliberately read a client's palm exactly opposite to what the lines on her palm suggested. The client told Hyman it was the most accurate reading she had ever had.

Schwartz's follow-up study addressed Hyman's complaints. It was, as Hyman himself described it, "admirably simple and well controlled." In this study, Schwartz tested one medium only—Laurie Campbell, from Los Angeles. Campbell traveled to the Human Energy Systems Laboratory, where she gave telephone readings for six unidentified sitters. Campbell's phone was on mute, so the sitters could not hear their readings; they simply sat quietly, as though on hold. The sitters were then mailed transcripts of two readings: their own (obviously not identified as such) and that of the other sitter called that day. The sitters were asked to rate the material in two ways. First, they were asked to rate each statement in the two readings as "hit," "miss," or "questionable." They were also asked to circle "dazzle shots"—described as statements that made them go, *Wow*.

Overall, the data were not dazzling. The percentage of hits among sitters' own readings was not significantly higher than among control readings. When Schwartz analyzed the dazzle-shot data, he reported "no evidence of anomalous information transfer." In other words, there was no statistically significant evidence that the medium was receiving information from the sitters' dead loved ones. This falls well in line with medium studies that have come before. "Where results turned out to be significant, they were often marginal and not impressive," concluded Schouten. "Even with a star subject . . . most experiments failed."

Schwartz did not consider the experiment a failure. He chose to focus on one reading in particular, that of a sitter named George Dalzell (who has since gone on to become a medium himself). Campbell's reading of Dalzell was judged—by Dalzell, though without knowing it was his—to be sixty percent accurate. Hyman suggested that Dalzell could simply be one of those sitters who tend to interpret most statements made by a medium as relevant or accurate. British psychiatrist and SPR officer Donald West did a two-year study of mediums and was amazed by the different rating styles of subjects. Some saw relevance in nearly everything the medium said, whereas others tended to dismiss almost all statements as inaccurate or meaningless.

Hyman complained that Schwartz's highlighting of just one reading reduces research to anecdote. But perhaps mediumship shouldn't be judged under the harsh lights of statistics and on-demand laboratory performance. If paranormal insights occur rarely, and largely outside of voluntary control, then perhaps it makes sense to focus on isolated moments—when the energy is right, whatever that might mean, and the medium is in fine fettle. But if this is the case, I think you have to wave good-bye to ever achieving "proof"—at least the kind that will stand up, statistically and methodologically, to the standards of peer review and academic orthodoxy. Medium research will become qualitative, not quantitative.

As is about to happen this afternoon, in Schwartz's lab. The discarnate who will be asked questions via Allison DuBois today is Monty Keen, a British colleague of Schwartz's who collapsed and died of a heart attack the previous month while arguing with a skeptic about telepathy at a panel in London. After DuBois finishes a blind reading for Keen—whose identity is not known to her and whom she has never met—and answers the formal afterlife questions for Beischel, Schwartz is going to chime in with some off-the-cuff postexperimental questions of his own. (Meaning, basically, that it won't be done in a manner that will satisfy Ray Hyman.) Schwartz is gathering material for a presentation he plans to make at an SPR tribute to Monty Keen being held in London the following month. He hopes to liven up his tribute speech with some appropriate quotes from Monty on the subject of life in the hereafter—from a resident's perspective. He also hopes to obtain so-called veridical information from DuBois—

statements that would prove, or at least strongly suggest, that it is truly Monty Keen communicating.

IF YOU REALLY wanted, as a discarnate being, to prove that it was you coming through—and not some product of a medium's imagination or subconscious—you could try to communicate the key to an encoded message. This would, of course, entail some setup while you're alive. You'd need to come up with a message, encode it, and then give the encoded gibberish to your friends and tell them what you're up to. After you die, you would then try to transmit the key word that decodes it. Of course, this assumes at least a passing familiarity with cryptography, and rare is the paranormal enthusiast who holds this sort of expertise.

Rare, but not unheard of. In 1946, Robert Thouless, a former president of the SPR, encoded two passages and published them in the society's *Proceedings*, along with an explanation of his intent and an invitation to encryption experts to try to break the codes while he was still alive, to ensure they were foolproof. A cryptographer quickly managed to decipher the first message, leading Thouless to seek the man's aid in coming up with an "unbreakable" code as an additional message. Thouless died in 1984, and the Thouless Project, headed up by emeritus University of Virginia professor Ian Stevenson, was born. Stevenson received some one hundred usable responses from people who had heard about the coded messages and had sat with a medium who believed he or she had discerned one or both of the keys. The keys were run through a computer program, along with an additional eighty variations on those responses. There is, of course, an avenue for fraud, as Thouless could have simply—were his intent not to solve the mystery of life after death, but to persuade people that it exists—told the key word to a medium before he died. Thouless addressed this in his article as best he could, stating, "I happen to be an honest man." I happen to believe him, for all decoded sequences have amounted to, to quote one Thouless Project report, "a meaningless jumble of letters."

Thouless himself had pointed out one obvious potential obstacle to the success of such tests, and that is that "the communicator is suffering from the disadvantage that he can no longer have the use of his material brain." Perhaps our memory disappears along with our neurons. Indeed, in 1986, honorary SPR secretary Mr. A. T. Oram reported to the society that he had made contact with Thouless via no less than eight different mediums and that "it seems he cannot remember the keys."

Low-tech renditions of this type of experiment have involved messages in sealed envelopes left in bank vaults and then opened after the message-writer's death and upon consulting mediums as to the content of the message. SPR cofounder F. W. H. Myers penned one such message, which British physicist and spiritualist Sir Oliver Lodge opened fourteen years after F. W. H.'s death, upon hearing from an overconfident medium in his acquaintance that she was certain she had received the message. The experiment, to quote Lodge's 1905 writeup in the SPR's journal, "completely failed."

As did later efforts to solve the Oliver Lodge Posthumous Test, consisting of a sealed packet (the Oliver Lodge Posthumous Packet) comprising seven envelopes each inside the next, a veritable filo dough of envelopes, with the late Sir Oliver's message inside the innermost. Following Lodge's death in 1940, a six-party committee (the Oliver Lodge Posthumous Test Committee) was convened and some 130 sittings with various mediums were held. Owing to the arcane complexities of Lodge's test, the proceedings quickly deteriorated into procedural bickering. All six outer envelopes held obscure clues intended to jog Lodge's memory but tending to give away aspects of the final message, which turned out to be a musical fragment tapped out on Lodge's fingers. You can't blame the mediums for being exasperated. Here, for instance, is the text of Sir Oliver's note in Envelope #3: "If I give a number of 5 digits it may be correct, but I may say something about 2 8 0 1 and that will mean that I am on the scent. It is not the real number . . . but it has some connection with it. In fact it is a factor of it." The flummoxed committee attempted to summarize four mediums' disappointing efforts, concluding weakly that it was best to "leave it to each reader to make his own evaluation." This reader's evaluation is that Sir Oliver was possibly a few envelopes short of a stationery set.

A macabre variation on the posthumous test was undertaken in 1921 by Thomas Lynn Bradford, variously described as an electrical engineer, a dramatic and humorous reciter, and a psychic investigator. (Bradford apparently excelled at none of these pursuits, for his entire estate at the time of his death consisted of five cents, some pawnshop tickets, three cheap watches, and several books on spiritualism.) According to a front-page story in the *New York Times*, Bradford had run an ad in a Detroit newspaper seeking to meet with others who were interested in the subject of "whether the dead can communicate with the living." A Mrs. Ruth Doran responded. The two met and agreed that, as the *Times* put it, "there was but one way to solve the mystery—two minds properly attuned, one of which must shed its earthly mantle."

Unlike Thouless and his ilk, Bradford was unwilling to wait around for natural causes to remove his mantle. He shed it that very night, by turning on the gas in his rented room. He was found that evening by his landlord, a Mr. Marcotte, who sat down with the police and told them what he knew about Bradford. This amounted to two colorfully disparate facts: "Mr. Marcotte says that in his younger days [Bradford] was a star athlete and was a champion pole vaulter of the Detroit Athletic Club. He would often put on a frock coat and impersonate Dr. Jekyll and Mr. Hyde." Though the juxtaposition is no doubt the result of hurried editing, I have chosen to believe that the impersonations were carried out while, rather than in addition to, pole vaulting.

The *Times* followed up with a much shorter piece the next day, headlined, "Dead Spiritualist Silent."

ALLISON DUBOIS IS seated in a striped armchair in one of the four rooms that make up the Human Energy Systems Laboratory. (At the time I visited, the lab was a converted two-bedroom stucco house owned by the university; it has since moved to a twenty-room suite near the medical school.) Schwartz and Beischel sit in matching armchairs behind DuBois, so their expressions can't be seen by the medium. On a wall is a placard with the Human Energy Systems Laboratory logo: a heart, a rainbow, and a human form with its arms outstretched, expressing love or joy or the size of the fish he caught, as people seem to do on New Age book covers. We've reached the postexperimental phase, where Schwartz is going to ask Keen, via DuBois, some questions of his own.

Schwartz's questions are paced and deliberate, adding an air of suspense to the room and making DuBois knit her brow. "Does this person . . . have . . . a message he'd like conveyed to his wife?" DuBois crosses and uncrosses her bone-china ankles. The air conditioner breathes.

"She's got new curtains."

Schwartz blinks in his seat. He's the boy who wanted trucks and got a chemistry set. "Mmm. Can he show us—"

DuBois cuts him off. Abruptly and with seeming certainty, she says: "He went down at a podium. . . . Like . . ." She snaps her fingers. "He falls, he goes down at a podium. Like an assembly. And he goes down at a podium. Is that odd? That's what he's showing. The man that died at the podium."

This is what Schwartz means by a dazzle shot. In my head, I did in fact go, *Wow*. Its dazzle is dimmed slightly by Schwartz having at one point asked DuBois whether the discarnate had "communicated to anyone else activities related to this research project," which affords a hint that the mystery entity might have been a fellow researcher. And if DuBois had heard about Keen's untimely and public death—which was the talk of the paranormal community a couple months back (even I had heard about it)—it might, at this point, have come to mind. Still, the suddenness and sureness of her statements don't jibe with a hunch or a fishing trip.

But what about all the statements that didn't seem to fit with Keen? Schwartz says it's possible that multiple discarnates come through when a medium opens up the channels; "crosstalk" is the official lingo they use. ("Fudge factor" is lingo I might use.) So it's up to the sitter, then, to pick out what's meant for him or her, and disregard the rest.

Or to interpret the statements more loosely. Take, as an example, some of the things DuBois said about my mother. Just before she began the Keen reading, DuBois stopped the proceedings to say that she was concerned that she might be getting crosstalk from my mother and from the father of a cameraman who was shooting the proceedings for a documentary. Both of whom are in fact dead. I later asked her what else she could tell me.

The reading was a more or less even mix of hits and

misses, with many of the hits being things that would fit a sizable percentage of the population (e.g., a cat in a sunny window, family gatherings being important). What I found most striking about the reading was the difference between my reactions to her statements and Schwartz's reactions. DuBois at one point said she was getting the letter *K* as being connected to my mother. "I don't know if she means the initial *K*, like Katherine or Kaye, but she's referencing *K* as connected to her," was what DuBois said. I drew a blank. My middle name is Catherine, but *K* seemed to be what the alleged discarnate was trying to put across. DuBois came back to the *K* again later in the reading. She seemed insistent. Finally I said, "Well, my middle name is Catherine, but . . ." Schwartz guffawed. "You are such a jerk! You expect it to be precise!"

But "Catherine" wasn't what came through, I protested. It was *K*. This sort of thing happened a few times. DuBois reported that my mother was making reference to "the man that still has the ring on his finger for her." This meant nothing to me, as my father never wore a wedding ring, or any ring at all. I mentioned that I wear her wedding ring, but since the statement was about a man, not a ring, I didn't think it was relevant. Here again, they felt I was being too picky. Was I? Was I being too literal, too negative? Or does Schwartz give his mediums too much latitude?

I was leaning toward the latter, when my own personal dazzle shot showed up. My reading was over and I was quibbling with Schwartz about something, when DuBois interjected, "I'm showing a metal hourglass, that you turn over. Does your brother have one?"

My brother collects hourglasses. (And here Schwartz nearly fell off his chair.) I was impressed, not only by its accuracy, but by its specificity and obscurity and by the sudden, assured declarativeness with which DuBois said it. It's hard to dismiss it. Yet it is equally hard, for me anyway, to dismiss all the statements that clearly did not fit my mother's life.

Here is the funny thing. Both Gary Schwartz and I believe ourselves to be neutral, unbiased inhabitors of the middle ground. I think Schwartz falls short of that territory, and he feels the same way about me. And he may be right about me. I am skeptical by nature.

I'm going to act contrary to my character and allow for the possibility that the other things DuBois was getting were, as they say, crosstalk. Maybe that *was* my mother coming through. But what's the meaning of it? Why would one of my brother's hourglasses be the image she chose to present to me? Was she simply trying to prove she was there? Then why not deliver my birth date or the name of our street or any of a thousand things that would more clearly suggest to me that it was her?

Maybe if I understood the mental processes of mediumship—the methods and limitations of spirit communication—I could answer these questions for us. This was what I was thinking when I committed the bold and ridiculous act of signing up for "Fundamentals of Mediumship," a three-day course at England's ancient and stately Arthur Findlay College.

7

Soul in a Dunce Cap

The author enrolls in medium school

ARTHUR FINDLAY was the president of the Spiritualists' National Union and a very wealthy man. Upon his death in 1964, Findlay bequeathed his enormous home, Stansted Hall, to the union to use as a college "for the advancement of psychic science." Accordingly, the upstairs portions of the building have been outfitted with two floors of dormitory-style bedrooms. These are arranged in six separate hallways, each portion of hallway reachable by one of two dozen possible combinations of interlinking stairways and obscured by fire doors and dead ends, such that a degree in psychic science is necessary simply to find one's bed in the evening.

"Fundamentals of Mediumship" begins on a Friday and runs through Sunday. We were instructed to arrive at 2 p.m., though nothing is scheduled until dinnertime. A staff person shows me to my room. I unpack my bag and then head back downstairs. Everyone seems to be in the gift shop, so I join them. People who enjoy fairies and dolphins have a wide range of purchase options here. I leaf through a copy of Findlay's gigantic opus *The Psychic Stream*, which should probably have been dammed at around page 200. For no good reason, I purchase an Arthur Findlay College box cutter and then wander upstairs to the museum. The featured exhibit is a display of paintings done "via precipitation" by mediums channeling artists in the spirit world. One painting is of Abraham Lincoln, who was said to have held séances in the White House, and one depicts Sir Arthur Conan Doyle "on a conducted tour of Hades."

I decide to get my coat for a walk around the grounds. My map includes evocative details such as "West Transept," but no schematic of the loco upstairs hallways. At last, sweating and cranky, I locate my room. My roommate has arrived and is propped on her bed, reading a romance novel. We make small talk for a minute, and then she says, "I foresaw that you'd be an American. Blond and chubby." I'm hardly blond or chubby, but I feel it's too early in our relationship to mention this.

The grounds are vast and beautiful and empty, except for a stone sundial that, if this weekend's weather is any indication, reflects a potent optimism on the part of the owners. I walk as far out as I can, in the direction of a single vigilant bull standing in the roughs at the edge of the property. On the way back, I am soaked "via precipitation."

The evening orientation is led by Glyn Edwards, the school's head tutor, which is what the English say in place of professor. Edwards recently suffered a neurological event that rendered his face uneven and his breathing deep and slurpy.

He has a huge head and an unsettling tendency to put it right up next to yours and confront you with a question. His hair is arranged in a swept-back country-and-western style, with sideburns that are allowed to spread out and roam the wide, pale plains of his cheeks. Over his turtleneck he has on a gilded medal strung on a ribbon, as though he'd recently taken first place at a swim meet. I find him unnerving and strange, but people here speak highly of him, at least as a medium. A sitting with Edwards is being raffled off at five pounds a ticket.

"Mediumship," he is saying, "is about proving that life after death is a fact. I challenge you to prove that what you feel this weekend is coming from beyond." Edwards introduces the five other tutors, each of whom will lead a group of us for the rest of the weekend. We are not an easy gathering to stereotype. Along with the more predictable assortment of New Agers and old spiritualists, there are trendy-looking Europeans from the continent, a clique of regular-looking English blokes, a Maltese retiree, a blind man.

"Is everyone a medium?" Edwards asks, but does not wait for an answer. "No. Can everyone become one? No." He looks around the room, breathing audibly. "It's about you finding out what you've got."

So far I've got: jet lag, an Arthur Findlay box cutter, a roommate who reads romance novels, a bad attitude.

Next morning we splinter off into our groups. My tutor is another popular English medium. She is reminiscent of Elizabeth Taylor in her forties: a rolling terrain of voluptuousness and eye shadow, balanced on tiny, dressy black heels. Mediumship came to her one night when she went to see a gypsy. She describes a curtain opening in her mind's eye, and suddenly there they were: the spit-its, to use her pronunciation. She says she was, until that moment, an atheist. Like Allison DuBois, she strikes me as intelligent and savvy. I want to get her drunk in the Arthur Findlay pub, take her aside, and say, "No, but really . . ." But I haven't the nerve. Like Edwards, she's an intimidating presence. You get the feeling they cut up doubters and serve them for lunch around here (all the more so once you've tried the lunches).

I've been very curious to find out how someone teaches a skill as ineffable and seemingly unteachable as spirit communication. Our tutor speaks to us for about fifteen minutes, but actual take-home instructions are thus far few. They amount more or less to this: Expand your energy. "Push out your energy, fill the room with your power." It seems to be something you just try to do. I try, I really do, but I have no idea where my energy is located or how to control its size or direction. I notice I'm moving my ears.

"Right," says the tutor after a minute has gone by. "Does anyone *not* feel a contact?" No one raises a hand. I haven't got my energy out the door, and apparently everyone else's is off in heaven at an ice-cream social. I raise my hand. The tutor comes over and puts her hand up to my face. She asks if I can feel my face. What does this mean? It's not numb, so I guess the answer is yes. I nod.

"Okay, good, you've got it." She turns back to the group. I don't read minds, but I think I know what's going on in hers: *AVOID THE YANK. The Yank is trouble.*

We're paired off to do our first reading. For now, we're told to try to pick up information about the person we're sitting with, rather than try to commune with his or her dead relatives. I'm with John, a soft-spoken, reticent man of maybe fifty, with a heavy Midlands accent. "Project your energy out to the person you're sitting with," our tutor tells us. "Encompass them. Get the feeling of them."

John looks unhappy. He doesn't look like he wants to be encompassed. "My wife brought me here," he confides. The tutor is making the rounds, so we have to do something. John rubs his face. He squints at me. "So we're supposed to *poosh* somethin' out?" He clenches his eyes shut. A minute passes. He opens his eyes. "I'm sorry, luv. I'm not gettin' *any*thin'." All around us, partners are jabbering enthusiastically. "You want to have a go?" John offers hopefully.

I tell him a boat comes to mind. His face is tan and lined, and he looks like a friend of mine who sails. You say these things not because you're trying to cheat, but just to be saying something, just to get on with it. John shakes his head. No boats. "Brown green striped wallpaper," I say next. "Big old homey sofa."

John leans forward in his chair. "*That* is incredible."

I'm not so sure it is. I imagine the wallpaper and sofa came to mind because John's accent sounds working-class and that's part of my image of working-class English living rooms.

The tutor has pulled up a chair beside the pair sitting next to John and me. The woman says she sees a den. Her partner nods. With a framed certificate, she adds. The man nods some more. "What color are the walls?" says our tutor. The woman says cream; the man says yes. "That's brilliant," the tutor says, getting up from her chair. "You quit while you're ahead."

We're learning, but what are we learning? Our tutor never said to us: Stick with the everyday. Try to be general because there's a better likelihood you'll be right. But we're picking it up anyway, or I am, at least. You want to get things right, because it's no fun not to. So you find yourself gravitating toward common, nonspecific attributes, things that apply to lots of folks—or dens, or what have you. No one is getting, say, the word "trilobite," or Jefferson Monument on a winter day, or the name Xavier P. Pennypacker. Because that would be a terrific long shot, and no one wants to set oneself up to be wrong. It's exciting to be right. Maybe you're psychic, you find yourself thinking, maybe you've made contact in spite of yourself. The little successes are their own reward.

I'm also learning to work from visual hints. Our tutor never told us to look at our partners' clothing and accessories to get a feel for their backgrounds, their milieus. But I found myself taking this tack with John right away, almost without thinking. And others are clearly doing this with me. Before we're done, three people will tell me they get the sense I'm a student. I'm not, but I am the only one taking notes in a notebook.

I am alone in my assessment of the class so far. When we get back together in a group, there's a buzz of excitement. I resolve to try harder.

After a lunch of soup and something ectoplasmic in tan sauce, we pair off again. This time we're to try to make contact with the spit-it realm, to pick up a transmission from someone dead who is known to our partner. I'm with Nigel this time. Nigel is upbeat and likable, and seems not the least bit self-conscious or confused about what he's been asked to do. He volunteers to go first. Straightaway, Nigel says he sees a man with a big belly and suspenders. "He's alone, he's drinking too much." He thinks it might be my father. My father did drink too much, but he never wore suspenders. It sounds more like a friend of mine who died of cirrhosis the week before I left. If you asked me to describe this man, that's pretty much what I'd say: big belly, suspenders, lonely, drank too much. It leaves me wondering. Men who drink too much often have big bellies, so there's nothing surprising there. It's the suspenders bit that grabs me, combined with the newness of this person's death. If I were inclined to easy persuasion, I might be persuaded.

Our tutor is behind us, working one-on-one with a serious young man named Alex. She's describing the house of a deceased grandmother. "I feel a problem with the windows."

He looks puzzled, shakes his head.

"Well, there is," our tutor insists. "Had she changed the curtains?" He shrugs. "Was she *thinking* of changing the curtains?"

This is something I've seen done by TV mediums. They seem to employ a subtle bullying. Blend this with the natural human tendency to want to please, and it's fairly easy to bring someone around to your point of view.

At dinner we sit at long tables, students from all the tutors' groups mixed together. The consensus seems to be that the course is fantastic. I do not meet a single person whose reactions or opinions resemble my own. Of course, a seminar like this self-selects for those prone to embracing New Age beliefs. It seems every other person I talk to has their Reiki 1 energy healing license. I'm the only person I can find who so far believes they have no mediumistic abilities. I am very much out of my element here. There are moments, listening to the conversations going on around me, when I feel I am going to lose my mind. Earlier today, I heard someone say the words, "I felt at one with the divine source of creation." Mary Roach on a conducted tour of Hades. I had to fight the urge to push back my chair and start screaming: *STAND BACK! ALL OF YOU! I'VE GOT AN ARTHUR FINDLAY BOX CUTTER!* Instead, I quietly excused myself and went to the bar, to commune with spirits I know how to relate to.

I have learned some things this weekend, though not what I came here for. I have learned that I was wrong about mediums. I no longer think they are intentionally duping their clients. I believe that they believe, honestly and with conviction, that they are getting information from paranormal sources. It's just a different interpretation of a set of facts. Mediums and the people who believe in them tend to, as the song goes, accentuate the positive. I tend to do the opposite. Maybe they're right. Maybe I am.

It seems to me that in many cases, psychics and mediums prosper not because they're intentionally fraudulent, but because their subjects are uncritical. The people who visit mediums and psychics are often strongly motivated or constitutionally inclined to believe that what is being said is relevant and meaningful with regard to them or a loved one. As Sybo Schouten put it, "It is the client who makes the psychic."

Nonetheless, there have been enough famously, flamboyantly fraudulent mediums over the decades that the men and women of the paranormal research community eventually began to look for ways to remove the middleman. Ways to communicate directly, person-to-person, as it were. If Alexander Graham Bell could make a disembodied voice hopscotch a continent, if Guglielmo Marconi could send invisible messages through the air from one town to the next, how hard could it be to forge a link with the Great Upstairs?

Can You Hear Me Now?

Telecommunicating with the dead

THE NATIONAL Forest Service has a fine and terribly dark sense of humor, or possibly they have none at all. For somebody, perhaps an entire committee, saw fit to erect a large wooden sign near the site where fourteen emigrants bound for California were eaten by other emigrants bound for California when they became trapped by the savage snows of 1846 and starved.* The sign reads: DONNER CAMP PICNIC GROUND.

---

*The Donner Party spent the winter of 1846–47 stranded near Donner Lake, in the Sierras of California. When it became clear there wasn't food to last the winter, seventeen of the strongest set out to get help. Another blizzard hit, stranding the rescue party at what came to be called the Camp of Death. The flesh and organs of four who died there—though not, I am relieved to report, the man named Mr. Burger—gave the others the strength to make it

I got here on a tour bus chartered by Dave Oester and Sharon Gill, founders of the International Ghost Hunters Society. IGHS, one of the world's largest (fourteen thousand members in seventy-eight countries) amateur paranormal investigation groups, sponsors ghost-hunting trips to famously and not-so-famously haunted sites. By and large, we look like any other tour group: the shorts, the flappy-sleeved tees, the marshmallow sneakers. We have cameras, we have camcorders. Unlike most visitors here today, we also have tape recorders. I am facing a pine tree, several feet from a raised wooden walkway that guides visitors through the site. I hold my tape recorder out in front of me, as though perhaps the tree were about to say something quotable. The other members of my group are scattered pell-mell in the fields and thickets, all holding out tape recorders. It's like a tornado touched down in the middle of a press conference.

A couple and their dog approach on the walkway. "Are you taping birdcalls?"

I answer yes, for two reasons. First, because, well, literally, we are. And because I feel silly saying, We are wanting to tape the spirit voices of the Donner Party.

Thousands of Americans and Europeans believe that tape recorders can capture the voices of people whose vocal cords long ago decomposed. They refer to these utterances as EVP: electronic voice phenomena. You can't hear the voices while you're recording; they show up mysteriously when the tape is replayed. If you do a web search on the initials EVP, you'll find

---

over the mountains. Lest you doubt the direness of the situation, a quote from *Unfortunate Emigrants*: "January 1, 1847. They made their New Year's dinner of the strings of their snow-shoes. Mr. Eddy also ate an old pair of moccasins." By the time help arrived, four months hence, most of those left alive had resorted to the food that knows no cookbook.

dozens of sites with hundreds of audio files of these recordings. Though some sound like clearly articulated words or whispers, many are garbled and echoey and mechanical-sounding. It is hard to imagine them coming from dead souls without significantly altering one's image of the hereafter. Heaven is supposed to have clouds and lots of white cloth and other excellent sound-absorbing materials. The heaven of these voices sounds like an airship hangar. They're very odd.

The EVP movement got its start in 1959, when a Swedish opera singer turned painter named Friedrich Jürgenson set up a microphone on the windowsill of his country home outside Stockholm, intent on recording bird songs. As Jürgenson tells it, a titmouse was suddenly and mysteriously drowned out by a male voice saying something about "bird songs at night." Soon thereafter, a man was heard humming "Volare."

At first, Jürgenson assumed he had picked up errant snatches of a radio broadcast. Tape recorder circuitries can indeed act as receivers, catching snippets of radio, CB, or walkie-talkie transmissions—especially if the transmitter is close by. He concluded that this was unlikely because, over the ensuing weeks, he picked up voices seeming to speak to him by name and, curiouser still, to his poodle Carino.

Jürgenson wrote a book, and the book caught the eye of a Latvian-speaking psychologist named Konstantin Raudive. Raudive picked up the EVP ball and ran with it. He ran and ran until he had seventy-thousand recorded "voice-texts" and a book deal of his own. The publishing of *Breakthrough: An Amazing Experiment in Electronic Communication with the Dead* in 1971 spawned a proliferation of do-it-yourself EVP societies, from Germany and the United States to Brazil, many of which still exist.

Unlike Jürgenson, Raudive didn't tape-record the air; he developed his own techniques. Often he taped radio static, the obnoxious hissing rasp between stations. Like Jürgenson, Raudive countered the possibility that he had recorded breakthrough radio broadcasts by pointing out that the voices spoke to him by name. And many, he said, spoke Latvian, though Raudive resided in Germany.

Around the time Raudive's book came out, a Cambridge University student named David Ellis proposed to investigate EVP as the subject of a two-year Perrott-Warrick Fellowship. For the past three days, I've been reading Ellis's book at the same time as I've been reading *Unfortunate Emigrants: Narratives of the Donner Party*, the latter having tainted my reading of the former, such that when Ellis refers to "disembodied entities," I have to stop and think about whether we're talking about souls or entrails.

As a parapsychology student, Ellis was more kindly inclined toward the research than, say, the average English or chemistry student might have been. I would hazard a guess that a student of most any other department might have rethought his fellowship topic upon encountering, for instance, Mr. G. A. Player, who believes the clicks and crackles of his old PTE radio to be manifestations of a disembodied female spirit. ("Mr. Player thinks she acts as a sort of capacitor," reports Ellis with admirably neutral tone.)

One of the first things Ellis did was to get Dr. Raudive and his kit inside a room screened to block radio transmissions. Though Ellis never believed radio broadcasts to be the primary source of EVP, it was something that needed to be controlled for. On several occasions, Raudive's recorded voices had been identified by others as having been part of broadcasts. What he interpreted as "I follow you tonight," for example, turned out to be a Radio Luxembourg announcer saying, "It's all for you tonight!" Raudive agreed to enter the screened room only once. No voices were recorded, though of course it's possible no discarnate entities were passing through the neighborhood. Ellis tried making his own recordings. He did get a few faint voices, but deemed the results neither encouraging nor conclusive.

My fellow ghost-hunter Rob Murakami is rewinding his tape recorder. A minute ago, I watched him step off the wooden walkway, walk to a cluster of trees, and stand for half a minute, his head bowed and his back to the trail, as though relieving himself amid the poplars. Murakami gives the impression of a man who enjoys life, no matter what life happens to be dishing up. His business card identifies him as the chiropractor of the Rose City Wildcats women's football team, suggesting that life routinely dishes up pretty enjoyable material. I'm guessing the trip was the idea of his girlfriend, who frequently feels ghosts "in the back of my throat, wanting to talk." Last night at Louis' Basque Corner, an entity in her throat dodged prime rib and potatoes to tell us that we "should have come when the melons are in season." (Based on the things people report them saying, ghosts strike me as quite senile, which I suppose is par for the course when you've been around two or three hundred years. Their tape-recorded vocalizations lean steeply in this direction. A selection from Raudive's collection of EVP utterances: "Please interrupt," "Might be Mary-bin," "Industrious!")

"Hm," says Rob. He puts his tape recorder up to my ear. "I got some odd thuds. Maybe I hit it by mistake, but I don't think so." He plays it again, this time for tour leader Dave Oester. I like Dave. He's a middle-aged minister of unspecified affiliation, with sloping shoulders and glasses that constantly slip down his nose. He has a big round torso and a head that seems to sit right on top of it, like a snowman's.

"Someone chopping wood," says Dave, smiling. Dave smiles every other sentence or so, not because something funny has been said, but just to keep things friendly. This morning, before we left, Dave played us a recording made from his first visit to Donner Camp. To me, it did not sound like communications of any sort, except possibly the sort exchanged between turkeys. I heard a rapid, metallic "gobba-lobba-ob." Dave heard: "I need more milk." One IGHS member said that, yup, she could hear it, too. Then again, during a dinner conversation earlier in the trip, this same woman heard "Siegfried and Roy" as "Sigmund Freud." The resulting image—Sigmund Freud with flowing hair and tigers and too much men's makeup—haunts me to this day.

Psychologists would nominate the "verbal transformation effect" as a possible explanation. B. F. Skinner once played nonsense sequences of vowels to subjects and asked them to tell him when they heard something with meaning. Not only did they hear words (with consonants), they were quite solidly convinced that their interpretations were correct.

The human mind is also adept at turning nothing at all into intelligible sounds. C. Maxwell Cede, an honorary secretary of London's Society for Psychical Research, described for David Ellis an experiment in which a group of people were handed paper and pencil and asked to help transcribe what they were told was a faint, poor-quality recording of a lecture. The subjects offered dozens of phrases and even whole sentences they'd managed to make out—though the tape contained nothing but white noise.*

---

*It's possible that the history of creatively interpreted white noise dates as far back as the Oracle at Delphi, where the priestess sat above a crack in the temple floor, below which could be heard the roiling waters of a spring. Dean Radin, senior scientist at the Institute of Noetic Sciences, has posited that the white-noise-like sounds of the water may have brought on auditory hallucinations. (The more common theory holds that ethylene fumes issuing from the spot were sponsoring the woman's altered state of mind. Ethylene—better known for making bananas ripe than for making priestesses bananas—can cause hallucinations in concentrated amounts.)

Konstantin Raudive seemed especially prone to the verbal transformation effect. At one point in Ellis's research, he had a group of people listen to purported utterances Raudive had collected and write down what they heard. Where Raudive heard "Lenin," others heard "glubboo," "buduloo," "vum vum," a bullfrog, a sudden change in tape tension, and "a low elephant call." Late in his career, Raudive became fixated on the vocalizations of a parakeet, which he believed to be channeling communications (in German) from the dead.

The most provocative recording to come out of Donner Camp this fine autumn day is a clear and relatively unambiguous whisper that turned up on the tape of a man named Charles. "Settings," it says. The less far-fetched explanation would hold that Charles at one point said something under his breath about changing the settings on his tape recorder, and then forgot that he'd said it. Charles insists he didn't say it, and while I believe him, it still seems more plausible than the alternative, which is that the soul of, say, George Donner manifested itself in Charles's tape recorder.

In the end, I would have to agree with Ellis's conclusions: "There is no reason to postulate anything but natural causes—indistinct fragments of radio transmissions, mechanical noises and unnoticed remarks—aided by imaginative guesswork and wishful thinking, to explain the 'voice phenomenon.' "

Ellis's conclusions are supported by the experiments of University of Western Ontario psychology professor Imants Baruss, published in the *Journal of Scientific Exploration.* Baruss is not a skeptic; quite the contrary. He told me he believes science has amassed solid evidence for life after death—in the form of research by Gary Schwartz (see Chapter 6) and Ian Stevenson (see Chapter 1)—but he does not consider EVP part of it. In eighty-one forty-five-minute tape recordings of radio static, he picked up the following: a low whistle, an occasional radio station breaking through, a squawking noise that "with imagination" might be a "hello," a truncated sound that one technician interpreted as her name (Gail), the sound of a kiss after Gail the technician said "hello," and a "Tell Peter," which Gail claimed sounded like a deceased woman she had known whose husband was named Peter. "While we have replicated EVP in the weak sense of finding voices on audiotapes," concluded Baruss, "none of the phenomena found . . . was clearly anomalous, let alone attributable to discarnate beings."

I'd buy that (and I might not employ Gail next time around), but I'm not surprised the EVP community took umbrage with the study: If the source of those few voices wasn't spirits, then what was it? I know it wasn't the task of the study to answer that question. Still, it does rather leave one twisting in the wind.

Are there other explanations for these odd snippets of voice? I contacted the German electronics giant Telefunken, because I'd read that they investigated EVP in the 1980s. I got a reply from Jürgen Graaff, who recently retired from the company after forty years as an engineer and, later, a managing director. He said he had heard of EVP, but did not know of any Telefunken-sponsored research. Then he told me about something called the ducting effect. Every now and then, strange goings-on in the electronic layers of the ionosphere create small "ducts" that enable fragments of radio broadcasts or walkie-talkie communications to travel thousands of miles. "A taxi driver communication in New York could suddenly be monitored for a couple of minutes in Europe," wrote Graaff in an e-mail. "From a classical engineering point of view, this ought not to be possible, as the power of a taxi transmitter is very small." Yet it happens. "After a few minutes, the ducts collapse and the phenomenon disappears. You can guess what I want to express about EVP!"

Talking with Graaff, it began to seem that the world of electronic broadcasting could serve up all manner of seemingly paranormal goings-on. Sometimes a gap between two pieces of metal, or a piece of metal and the ground, can set up a sparking that serves to demodulate a radio signal if a transmission is especially powerful or the tower close by. Graaff recalls a hysterical East German woman whose roasting oven, she said, would speak to her whenever she opened the door. A man who lived in the same neighborhood was being addressed nightly by his heating system. Engineers dispatched to look into the reports identified the words as segments of the nightly Broadcasting in the American Sector broadcast and reassured the shaken citizens.

Graaff thereby confirmed something I'd long assumed was an urban myth: that dental fillings can pick up radio transmissions. Perhaps you recall the episode of *The Partridge Family* wherein Susan Dey announces that she can hear the Rolling Stones in her mouth. The show implied that the music is so clear that if David Cassidy were to put his ear right up to your mouth—close to but not quite my sixth-grade fantasy—he could name the song. Graaff explained that if two metals are used side by side—say, an old amalgam covered by a gold cap (or, in Miss Dey's case, braces and a filling)—a small gap between them can foster what's called a semiconductivity effect. A jumble of low tones could indeed be heard, though probably only as far as your own inner ear, meaning that Mr. Cassidy would have to work his head clear inside your eustachian tube.

I asked Graaff whether any of the Germans had interpreted their appliances' words as dispatches from the Beyond. He told the tale of a farmer who owned the fields around the mighty Elmshorn transmitting station where Graaff used to work, just north of Hamburg. "He'd been walking the fields, checking the fences, when all of a sudden he came running to the station manager, deadly pale, saying, 'Sir, I heard the Holy Ghost speaking to me! It came from a piece of wire sticking out of the ground!'" The Ghost spoke in the same cryptic, truncated manner effected by Raudive's and Jürgenson's ghosts. Graaff and the manager followed the farmer out to the wire, which was whispering and hissing when they arrived, and every now and again issuing an intelligible phrase. The manager leaned down and pulled the wire from the earth, silencing the Holy Ghost and leaving the farmer to more pedestrian concerns, like the effects of two-hundred-thousand-watt radio towers on farm animals.

You can see and hear your own Holy Ghost if you visit the grounds of an exceptionally robust transmitter, such as the ones operated by Voice of America. Wander up to the metal fencing around the facility after dark, Graaff says, and you might be able to see pale glimmering sparks here and there along the metal. Lean in close and you may hear the sparks singing—or talking, depending on what's being broadcast.

WILSON VAN DUSEN was the chief psychologist at the Mendocino State Hospital in northern California for many years. This was an inpatient facility for the severely mentally ill—chronic schizophrenics and alcoholics, the brain-damaged, the senile—so he spent a lot of time listening to his patients talk about their "others": the voices in their heads who cussed at them and threatened them and needled and harassed them—or, very occasionally, encouraged and inspired them. At one point, he decided to try to talk to the voices themselves. "I would question these other persons directly," he wrote in a pamphlet entitled *The Presence of Spirits in Madness*, "and instructed the patient to give a word-for-word account of what the voices answered. In this way, I could hold long dialogues with a patient's hallucinations." At one point, he was administering Rorschach inkblot tests to the voices. I began to picture the hallucinations as actual inpatients, scowling men in ratty slippers, muttering in the corridors and disrupting bingo games. After interviewing twenty such patients, he decided that he agreed with the patients that their "others" were not hallucinations but inhabitants of a different order of beings.

Dr. Van Dusen is a Swedenborgian—a follower of the teachings of Emanuel Swedenborg, an eighteenth-century mining engineer/inventor/anatomist who began having religious visions in his forties. Swedenborg gained renown as a philosopher and wrote at length about the heaven of his visions, a dream realm inhabited by wingless angels and demons, which, he held, had once been mortal humans. Van Dusen began to notice that his patients' "others" fell into similar camps of good and evil, with the evil well outnumbering the good, and that they shared numerous traits with Swedenborg's opposing spirit entities.

You might be thinking, and I could not blame you, that it is more plausible that Emanuel Swedenborg was having schizophrenic episodes than that the schizophrenics were having Swedenborgian episodes. However, by all measures, Swedenborg was not psychotic. He maintained a productive existence as a statesman and theologian, and enough people took—and take—him seriously for the Swedenborgian Church to have become, and to remain, a thriving international denomination.

I was introduced to Van Dusen's theories by an EVP

enthusiast who was thinking of investigating the possibility that the voices schizophrenics hear are the same voices that wind up on EVP tapes, i.e., voices of discarnate entities. I ran this by the folks at England's Hearing Voices Network, a support organization for people with auditory hallucinations. My e-mail was answered by a helpful and forthcoming staffer and "voice-hearer" named Mickey who said that although it is network policy to accept all members' explanations for their voices, and although he didn't know much about EVP, it was his personal opinion that the notion was nonsense. However, he did know quite a few people whose voices seemed so real to them that they tried to tape-record them. The voice-hearers inhabit the opposite conundrum of the EVP people: The voices are audible (to them) at the time, but the tapes are always blank.

Thomas Watson, coinventor of the telephone, describes in his autobiography being contacted on several occasions by schizophrenics who believed that the words in their heads were being secretly broadcast from distant individuals. Most sought his advice on how to block the signals, but one enterprising psychotic offered—for a fee of fifteen dollars a week—to let Watson "take off the top of his skull and study the mechanism at work":

> He told me in a matter-of-fact tone that two prominent New York men . . . had managed surreptitiously to get his brain so connected with their circuit that they could talk with him at any hour of the day or night wherever he was and make all sorts of fiendish suggestions. . . . He didn't know just how they did it, but their whole apparatus was inside his head. . . . I excused myself from starting to dissect him at once on the grounds of a pressing engagement.

Mickey directed me to the research section of the Hearing Voices Network website, where it said that if a brain scan is done on a schizophrenic as he or she is hearing voices, the scan will show activity in the part of the brain involved in speech production. Meaning that the voices are the "inner speech" of the person who hears them.

BOTH JÜRGENSON AND Raudive have long since moved on to the other side of the tape recorder.* (David Ellis wrote Raudive's obituary in the *Journal of the Society for Psychical Research*, noting, in a classic *JSPR* moment, that the "strain of a conference on the parakeet voices . . . proved too much for him.") Their deaths did not extinguish the worldwide enthusiasm for EVP, nor did David Ellis's fellowship findings. In skimming the newsletters of EVP groups, I find the phenomenon treated ipso facto as communication with the dead. Why, given the negative findings of respectable, open-minded academics, are these folks so certain?

"It's one thing to get enough evidence to convince yourself, but it's a whole other matter to produce a demonstration that would be acceptable to a community of scientists," says Imants Baruss. Dean Radin, a former electrical engineer and the senior scientist at the Institute of Noetic Sciences in Petaluma, California, agrees. "EVP researchers may be genuinely sincere, but insufficiently critical to assess their own results." They're convinced by what they've heard, and that is enough.

---

*Literally, upon occasion: EVP literature holds that Jürgenson has had cameos on the tape recordings of an Italian EVP enthusiast, while Konstantin Raudive has made repeated appearances in the static on the TV screen of a couple in Luxembourg.

The sun is packing to leave when Dave Oester joins me on the walkway. I tell him I'm not getting anything. He asks me if I introduced myself to the entities before I started taping. "That's important," he says. "I always say, 'I'm Dave Oester of the IGHS, and I'd like to document the existence of life after death. I'd like to get your permission.' "

I clear my throat. "HI, I'M MARY ROACH . . ." You can't see where these guys are, so it's hard to know how loud to talk. "I'M WITH THE IGHS, THOUGH NOT ACTUALLY A MEMBER AS SUCH."

"You can say it to yourself, Mary. They read your thoughts."

"They do?"

Dave nods his head. "Sure they do."

Well, no wonder they're ignoring me.

THROUGHOUT HISTORY, each new breakthrough in the science of communications is inevitably recruited by someone with a shining for things spiritual. As magicians like Houdini and Britain's Harry Price began exposing the elaborate parlor tricks of the spiritualist mediums, promoters of the afterlife began incorporating gadgetry into their routines. Machines lent an air of scientific respectability to their claims. They promised a purer, seemingly less corruptible connection with the dead. You can't trust a human not to fake ectoplasm out of sheep lungs, but you ought to be able to trust a machine.

So instead of a medium speaking in a trance, you had a medium operating a "psychic" typewriter or Morse code console or Vandermeulen Spirit Indicator. You hadn't eliminated the middlemen, you'd simply outfitted them with impressive-looking machines. Séances were more technically complicated, but fundamentally unchanged.

Recording devices proved immediately popular with the spiritualist mediums—not to pick up otherwise inaudible communications, but to bolster believability. For what is a recording but a means of capturing and preserving something otherwise fleeting and unprovable? I think Dr. Neville Whymant put it best. An eminent—and eminently corruptible—scholar of the Orient, Dr. Whymant had been called upon by his friend Judge Cannon to speak for the authenticity of a phonograph recording of Confucius, speaking through the voice of medium George Valiantine, at a séance in Cannon's home in 1926. Valiantine was said by Whymant to be speaking in a (conveniently) "extinct" Chinese dialect. "I think you will agree," observed Whymant, "that though it is possible that you might hallucinate people, you could not hallucinate a gramophone."

Phonograph historian D. H. Mason spent weeks trying to track down a copy of the Confucius sessions. He did not succeed, though he did manage to find an itemized description of a boxed set of Valiantine recordings. Highlights included a war whoop by Valiantine's main spirit guide Kokoan, and "a pathetic song" sung in a shrill falsetto by Bert Everett, another Valiantine guide.*

---

*A note about spirit guides. You will occasionally read piffle about differences between the EEG of a medium and that of her guide, or control, and how this proves the guide's existence as a separate entity. In 1981, Gary Heseltine, now an epidemiologist with the Texas Department of Public Health, experimented with the EEGs of two unnamed mediums and their spirit guides Shaolin and Monsanto (the "Comanche chief," not the fertilizer concern). Heseltine writes that since sensory and metabolic input affect EEGs, you

Mason published a three-part article, including discography, on the topic of séance recording sessions. While the early efforts were merely recorded documents of the sittings—one particularly vigorous medium held forth sufficiently long to fill nine twelve-inch double-sided 78s—very soon the mediums took to singing while in trance, in the persona and voice of the spirit guide. Not surprisingly, given the preponderance of female mediums, the spirit guides (most of them male) tended to be tenors. It was an odd coupling: the high, sweet tones of the tenor register issuing from entities with hypermasculine handles like Power or Hotep. Perhaps this explains the appearance, in 1930, of an Italian spirit guide. Sabbatini, the Italian tenor, began turning up at the séances of prominent Cape Town medium Mrs. T. H. Butters. Mason quotes a description of a Sabbatini performance in a 1931 issue of *The Two Worlds*, the newspaper of the Spiritualists' National Union: "While Mrs. Butters was under the control of the spirit, he delighted the sitters by singing Italian songs in a ringing tenor voice, and so powerful were the manifestations that in March this year the friends of Mrs. Butters decided to make a gramophone record of the voice." The recording quality was diminished somewhat by Mrs. Butters's tendency to stray from the microphone and move about the room "making operatic gestures," but was otherwise deemed to be of excellent quality.

This obscure musical genre reached its peak on April 3, 1939, when London's Balham Psychic Research Society held a séance *inside the studios of the Decca Record Company.* Presaging

---

would have to go to the extreme of "paralyzing and maintaining the medium on life support" to control these factors. Even then, he doubted you'd have proof. "Short of a high brain stem transection," Heseltine concluded, "it is difficult to conclude that differences in the EEG cannot be a consequence of differing sensory inputs."

the current vogue of single-name recording artists, our singing spirit guide was billed simply as Reuben. Reuben, performing via the vocal cords of medium Jack Webber, entertained séance guests with baritone renderings of "Lead Kindly Light" and "There's a Land," an anthem made famous by renowned English contralto Madame Clara Butt.*

Whether the spirits sang or simply spoke, the new recording technologies expanded the medium's options for income. In addition to holding séances, he or she could also sell tapes or records. The largest "direct voice"—meaning no spirit guide was employed; the deceased spoke directly through the medium's voice—recording enterprise was that of British medium Leslie Flint. Flint, who died in 1994, managed to attract a highbrow crowd of discarnates to his séances. If you run a web search on him, you'll find sites where you can hear lengthy postmortem recordings of Gandhi, Oscar Wilde, Chopin (who has, we learn, resumed composing following a brief stint decomposing), the Archbishop of Canterbury, and renowned Shakespearean actress Dame Ellen Terry. (More on Dame Ellen later.)

As was the custom, Flint carried out his séances in darkness. He insisted that the voices came not from his own voice box but from one built up, to quote one website, "from ectoplasm drawn from the medium and the sitters." The site displays a circa 1960 photograph of Flint, seated calmly in a chair, wearing suit, tie, horn-rimmed glasses, and what appears to be

*Oh, for the days when a nation's highest-paid recording star could be a beefy six-foot-two oysterman's daughter named Clara Butt. So remarkable was her voice that Madame Butt, as she was known early on, was recruited at a tender age to sing private concerts for Queen Victoria. Her lauded career in opera paved the way for what must have been a much-welcomed shift in titularity to Dame Clara.

the aftermath of a cafeteria food fight. The caption says, "Flint with ectoplasm resting on his shoulder." I don't know what Gandhi or Chopin sounded like while alive, so I can't comment on the verisimilitude of the recordings. But I can comment that Leslie Flint said he discovered his gift one evening at the cinema, upon hearing—as one will at cinemas—"voices whisper in the dark."

It was not only the mediums who were fond of gadgetry, but the paranormal researchers who put them through their paces. Initially, technology was recruited to prevent fraud or, more often, to document or quantify the mediums' powers. With few exceptions, the devices were christened in the syllabically overwrought vernacular of the Serious Laboratory Device. Microscopes now had to share the lab bench with Dynamoscopes and Telekinetoscopes. The staid and stately Ometer family, heretofore limited to Thermo, Baro, Speedo, and Sphygno, was asked to take in the Sthenometer, the Biometer, the Suggestometer, the Magnetometer, and the Galvanometer.* I tried to track down even one of these machines, but, oddly and disappointingly, no museum or private collection seems to exist. "The psychical organizations didn't approach these things from a historian's perspective,"

*The Ometer decline has continued, largely at the hands of the textile industry, who have given us the FadeOmeter, the Crackometer, and the Launder-Ometer (not to mention the Atlas Perspiration Tester, the Shirley Stiffness Tester, and the Evenness Tester 3 With Hairiness Module). Further Ometer abuse comes from the Centers for Disease Control (the Flu-O-Meter), the Royal Society for the Protection of Birds—their Splatometer tracks the abundance of flying insects, whose decline spells trouble for birds—and Gary Ometer, former Director of Debt Management for the U.S. Department of the Treasury. I was hesitant to phone Gary, for his title led me to expect a man of, shall we say, high scores on the Shirley Stiffness Tester, but he was a good sport about it. Gary blames shabby Ellis Island bookkeeping for his family's contribution to the Ometer situation.

says Grady Hendrix, former office manager of the American Society for Psychical Research in New York City. "These gadgets weren't something that more modern parapsychologists would have deemed worthy of saving. It's not an era they're proud of."

As the reputations of mediums continued to erode, paranormal researchers turned their attention toward devising some sort of direct spirit communication device, something that would remove the medium from the process entirely. F. R. Melton's Psychic Telephone managed to get the medium out of the room, but not entirely out of the picture. The "telephone" consisted of an inflatable rubber bladder attached to a transmitter, attached, in turn, to a pair of headphones. The bladder was said to contain "psychic air" full of spirit voices that could be amplified and transmitted into the headphones. How do you fill a balloon with psychic air? You have a medium blow it up. Magician-cum-paranormalist Harry Price tested the device in his National Laboratory for Psychical Research and found it to be, literally and figuratively, so much hot air.

THE LURE OF the gizmo remains strong among modern-day paranormal hobbyists. This is evident here this morning in Assembly Room A of the Golden Phoenix Hotel, where our group has gathered for Dave Oester's morning lecture. It is a standard hotel conference room, with long folding tables and a wooden speaker's pulpit and the blenderized teals and mauves of institutional carpeting. While they wait for Dave and sip at cups of coffee, my fellow enrollees trade tips about their kit, which they have spread out in front of them on the tables: meters, compasses, cameras, recorders. Spirits rarely register on human sensory systems, but, the thinking goes, that doesn't mean they don't exist.

People in ghost-busting groups posit that the reason humans can't normally see or hear the dead is that they exist in and communicate via the far extremes of the visual and auditory spectrums: light waves we can't see and sound waves we can't hear. This is why ghost-hunting groups use cameras with film that is sensitive to infrared rays, and why Dave Oester used to carry around a bat detector. He reasons that perhaps the dead, like bats, emit sounds in the ultrasound range. When I got home, I called Bill Gerosa, the president of the company that makes a device called the Belfry Bat Detector. I told him that ghost-hunters say his device can be used to detect spirit communications. "I can neither support nor refute that statement," said Bill after a few moments of quiet. He went on to say that not just bats, but rodents, insects, TV sets, and car brakes emit ultrasound, so there's a distinct possibility that the entities communicating via bat detectors are katydids or Chevrolets.

The woman seated beside me is fiddling with a handheld meter of some sort. She has the instruction manual open. A heading says, "ELF RESEARCH IN THE 90s." I like this woman, and I don't want to think the things I'm now forced to think about her. I ask her if she has ever seen an elf.

She stares at me suspiciously, like she doesn't need a Belfry Bat Detector; she can just *see* them flying around in there. "Nooo-o . . . Why, have you?"

I squint at the copy. "You can't see, smell, or touch them," it says, "but they are present in your everyday life." I am working on the phrasing of my next question when her boyfriend leans forward. "*E-L-F*," he says. "Not 'elf.' It stands for Extremely Low Frequency." As in background radiation. As in microwave ovens and overhead power lines.

Aha.

Nearly everyone in our group has brought along either an ELF meter or an EMF (for measuring electromagnetic fields) meter. The link between electricity and spirits is a tenacious one, and it dates back some hundred and fifty years. Standing on Oester's sloping shoulders are no lesser dignitaries than Thomas Edison, Nikola Tesla, Alexander Graham Bell, and Bell's partner Thomas Watson.

What you need to know is that the heyday of spiritualism—with its séances and spirit communications zinging through the ether—coincided with the dawn of the electric age. The generation that so readily embraced spiritualism was the same generation that had been asked to accept such seeming witchery as electricity, telegraphy, radio waves, and telephonic communications—disembodied voices mysteriously traveling through space and emerging from a "receiver" hundreds of miles distant. (Bell and Watson's claims for their telephone were initially greeted with more hooting skepticism than were the mediums' séance shenanigans; like Edison,* they took to touring the country with their gizmos, doing public demonstrations.) Viewed in this context, the one unfathomable phenomenon must have seemed no more unbelievable than the other.

Electromagnetic impulses seemed to provide the missing

---

*The publicity stunt is one of the lesser-known Edison inventions. In 1903, as part of a scheme to discredit the alternating current system (Edison was a DC man), he got involved in the Topsy-the-elephant situation. The Coney Island pachyderm had been sentenced to die for having killed three of her handlers. (One fed her a lit cigarette, so in my mind the jury's still out.) The swift and humane execution of an elephant was proving troublesome. Cyanide had failed, and hanging promised all manner of logistical turmoil. Edison called the ASPCA and suggested electrocution. He filmed the highly effective dispatch, and used it as proof of the dangers of AC.

explanation for—the absent science behind—mediumistic communication. If one accepted the workings of the radio and the telephone, spiritualism didn't seem like such an enormous leap. These devices must have made it seem much more plausible that, as Gavin Weightman writes in *Signor Marconi's Magic Box*, "individuals with special powers really could act as 'receivers' of invisible and inaudible signals." Weightman adds that staunch spiritualist Sir Arthur Conan Doyle would talk about how the greater distances traveled by nighttime telegraphic impulses were proof of "the mysterious 'powers of darkness' which spiritualist mediums exploited."

Even the inventors themselves viewed the etheric and the electric with the same set of awe-fogged eyes. Electricity maverick Michael Faraday, in writing about his experiments with electric eels, marveled that the work was "upon the threshold of what man is permitted to know of this world." Thomas Watson, in his autobiography, referred to electricity as an occult force. Like a surprising number of his peers in the scientific community, Watson dabbled in spiritualism. He spent two years believing he had a halo.* At one point, when prototype telephones were failing to reliably deliver coherent sentences, Watson endeavored—via a medium and without telling Bell—to ask those age-old experts in breakthrough communications: the dead. (The suggestions, alas, were "rubbish.") Watson was constitutionally prone to thinking outside the box—nay, several counties distant from the box. In fine-

*Until he figured out that his "halo" was a reflection of sunlight at a certain angle, Watson believed himself to have been singled out for some great purpose. "I told my mother about my halo," he writes. While Mrs. Watson did not come right out and say she could see it, she did the motherly thing and said it "didn't seem at all strange to her that her son was thus distinguished." Emboldened, Watson confided in Alexander Graham Bell. Bell told him to get his eyes examined.

tuning the speech transmission qualities of the nascent telephone, he tried out diaphragms—the part that vibrates when a caller speaks—of varying shapes and materials, including a human eardrum and bones. Watson borrowed the item from Alexander Graham Bell (*Al, lend me your ear!*), an authority on the mechanics of human speech. Bell got the ear from the aurist Clarence Blake, who got it from one of his patients after, Watson is careful to point out, he "had finished with it."

Watson's spiritualist beliefs colored his views of science, and vice versa. He viewed mediums as people with special powers to transform bodily radiations into a mechanical force, much the way a telegraph transforms pulses of electricity into audible bursts of Morse code. The science of electromagnetic forces offered a logic for the highly illogical rappings and table tiltings of the séance circle. Andrew Cooke, a Royal College of Art student whose insightful master's dissertation popped up in one of my web searches, wondered whether spiritualism inspired, rather than simply influenced, the minds of some of the great inventors. Of course the invention of the telegraph prompted mediums to begin taking Morse code dictations from spirits during séances. No surprise there. More intriguing is the inverse possibility: that the coded raps of early mediums like the Fox sisters sparked the idea for long-distance communication via Morse code.

Watson's faith in mediums was unique among the great electricians. Edison, Tesla, and Bell believed that the soul survived death and traveled, like a wireless impulse, to some etheric realm, but they did not, in the end, buy into the mediums' claims. (As Edison put it in his diary, "Why should personalities in another existence or sphere waste their time . . . play[ing] pranks with a table?") If anyone was going to make reliable, intelligible contact with the dead, they believed, it was inventors like themselves. Bell and his brother signed a pact to the effect that whoever died first would attempt to make contact with the other through a more reliable channel than the séance medium.

Tesla was a special case. He was, by his own description, exceptionally sensitive. "I could hear the ticking of a watch with three rooms between me and the time piece," he wrote in his journal. "A fly alighting on a table in the room causes a dull thud in my ear." Around the time his mother died, Tesla, under the sway of his mentor Sir William Crookes—famous for making rarefied gas glow green in vacuum tubes and infamous for thinking it was ectoplasm—tried to turn his antennae toward the paranormal. One night when his mother was on her deathbed, he slept with "every fiber in my brain . . . strained in expectancy." Early that morning she did indeed die. He recalled a dream of an angel with his mother's features, though he ultimately decided that the dream face matched a painting he had recently studied and that that explained "everything satisfactorily in conformity with scientific facts." Despite a fascination with the mysteries of death, Tesla did not, as far as I know, try to build a device for postmortem communications.

But Thomas Edison did. He describes in his *Diary and Sundry Observations* being engaged in the design of an apparatus that would enable "personalities which have left this earth to communicate with us." Edison imagined living beings as temporary conglomerations—"swarms" was the word he used—of infinitesimally small "life units" that persisted after death in a kind of loitering, dispersed form, and eventually regrouped as someone or something else. He described his machine as a sort of megaphone. He reasoned that the "physical power possessed by those in the next life must be extremely slight," and that, like the speck-sized Whos in *Horton Hears a Who*, they require a certain level of amplification to make themselves heard. Sadly, Edison himself departed for Whoville before completing the contraption.

Perhaps because of his amplifier project, Thomas Edison is often credited with the invention of something called the Psycho-Phone. Dave Oester says that the Psycho-Phone is the inspiration behind the ultrasonic transceiver he himself, along with an electrical engineer he knows, has been working on. Their device, which Oester hopes will facilitate two-way real-time communication with the dead, is called the TEC, short for Thomas Edison Communicator.

I looked into the Psycho-Phone, thinking perhaps Edison had a plan B. According to Tim Fabrizio's and George Paul's *Discovering Antique Phonographs*, the Psycho-Phone did indeed exist, but it wasn't designed for paranormal communications. It was an early, phonographic precursor to the modern-day subliminal self-improvement tape. As with the tapes, the listener sets the device to go off while he or she slumbers, in the hope that he or she will, say, to use an actual Psycho-Phone example, "wake refreshed—invigorated—and enjoy a regular bowel movement."

In 2003, a Psycho-Phone was put up for auction. There was a website posting written by the winning bidder, who believed that she had come into possession of one of Edison's devices. The woman was understandably puzzled by a transcript of one of the subliminal messages, which she found in the box and took to be a letter from "someone that perhaps was a little deranged":

> I enjoy drinking clean water or clean water flavored with the juices of pure fruit. Every morning I will get up in time to do a series of exercises to strengthen my body. My scalp is getting healthier every day as the blood flows abundantly. . . . My hair is growing luxu-

> riantly dark and beautiful. My scalp is glowing with health and new beautiful hair is growing thereon. I am a good mixer & have a wonderful memory.

These days, electricity, radio waves, and telephonics are the stuff of everyday life on earth. They've left the realm of the mysterious, and in their place we have ultrasound, infrared, cyberspace. Ultrasound was the mystery force *du jour* among paranormalists long before Dave Oester began tinkering with it. In the 1980s, an electronics buff named William O'Neil, who had a taste for the paranormal and a lab full of oscillators and ultrasonic receivers, developed Spiricom, a device for spirit communication. He claimed to be having lengthy two-way conversations with the dead, or anyway one of them: a former NASA physicist named George Mueller. O'Neil and fellow paranormalist George Meek published hundreds of pages of transcripts of the Spiricom conversations, including lots and lots of shop talk:

> *(Dead guy) Mueller*: By the way, did you get that multi-faceted crystal?
> *O'Neil:* No, I got that five-faceted one from Edmund's.
> *Mueller:* Edmunds? Who is Edmunds?
> *O'Neil:* Edmund's is a company. Edmund's Scientific.
> *Mueller:* Oh, I see. Well, very good.

The dead are surprisingly poor conversationalists, given all the novel and mind-blowing things going on in their lives. They're like ham radio operators. I once stumbled onto a long series of ham radio transcripts on the web. Here you'd have these two men, say, a Minnesotan and a Vanuatan, speaking to each other from what may as well be different planets, and they can't think of anything to talk about but their equipment. "What are you using there, a KW-50?" "Oh, no, I like a Hammarlund. Got the Q multiplier built right in."

Dull as their man was, the concept of a chat with a dead man was pretty darn exciting, and Meek made the highly questionable decision to go public. In 1982, he booked the National Press Club ballroom in Washington, D.C., and sent out a press release announcing "electronic proof that the mind, memory banks and personality survive death." A *Chicago Sun-Times* reporter expressed disappointment that, owing to technical difficulties, no "live" demonstration was possible. To make up for it, Meek played a tape of what he claimed were the astral voices—presumably obtained via Spiricom—of newspaper mogul William Randolph Hearst and . . . the great Shakespearean actress Dame Ellen Terry.

Unlike Mr. Flint, Meek and O'Neil had no apparent plans to profit from their project. They let someone else write the book, and they handed out blueprints for the device at the press conference, encouraging others to try to replicate what they'd done. (No one succeeded.) If it was a hoax, it was a perplexing one. Way too much work for questionable payoff. If it wasn't a hoax, it was . . . what? Real? There seemed to be no possible middle ground. I asked Dean Radin what he thought about it. "The middle ground between genuinely true and outright faking," he offered, "is unconscious delusion."

LIKE MANY OLD structures in England, Staffordshire's Westwood Hall has a long-standing reputation for being haunted. In 1998, the school's caretaker was preparing a paper on the history of one Lady Prudentia Trentham, who died on the grounds and is thought to be the source of the alleged haunting. When the caretaker spell-checked his paper, strange things began to happen. For instance, when the program highlighted the misspelling "Prudentiaa," it did not offer "Prudentia" as the proper spelling. The spellings it suggested were: "dead," "buried," and "cellar." This sort of thing didn't happen when he spell-checked other documents, and it happened on two different computers.

Thinking that the discarnate Lady Prudentia was trying to communicate with him via his spell-checker,* the caretaker called the Society for Psychical Research. The SPR took the claim seriously. This wasn't, after all, the first purported instance of dead spirits using computers to communicate. Far from it. Spirits have also, if you buy into the literature on "instrumented transcommunication" (a close cousin of EVP), made use of TVs, VCRs, alarm clocks, and answering machines.

In 2001, the team of SPR researchers who were looking into the case hired software consultants Julie and David Rousseau to come take a look at the computer and the software to see if perhaps the caretaker's system had been hacked into by capricious spirits still of the flesh. The Rousseaus confirmed that an experienced programmer could, without too much trouble, create the effects that the caretaker had seen. But a program like this would be simple to detect, and they quickly determined there'd been no foul play.

This left two possibilities: a bug in the software or a ghost in the machine. To test for the former, they attempted to recreate the phenomena on a document and computer of their

*I can easily relate to the feeling that one's spell-checker is possessed. Mine recently informed me that "fucking" is not a word, but that "cucking," "rucking," and "funking" were all good words that I might like to substitute.

own, using all the same steps and software (Microsoft Word 6) that the caretaker had been using. They soon succeeded. (It is worth pointing out that Julie Rousseau is open to the possibility that paranormal forces can influence computers. She serves on the council of the SPR.)

The bug involved the custom dictionaries that the caretaker had set up and the unorthodox manner in which he had modified them. Our man had noticed that the peculiar spell-check offerings always seemed to involve words from one of his custom dictionaries. Because he believed the spell-check anomalies to be communications from Lady Prudentia, he had decided to expand her vocabulary by seeding his custom dictionary with dozens of ordinary words—rather than simply the proper nouns and place names related to the file.

Rousseau found that when she used Word 6 on her computer, the bug commenced on the twenty-first misspelling of custom dictionary words. The anomalous offering, she figured out, is simply the word that was last taken from the custom dictionary as an alternative suggestion for a misspelled word. Because the caretaker had four custom dictionaries operating, the bug kicked in much sooner than it would otherwise—which helped explain why the bug hadn't been reported by other users.

You would think that the matter would have come to rest there, but it has not. The caretaker insists that some of the anomalous suggestions are *not* words he added to his custom dictionaries. Again, the Rousseaus offered a possible explanation. The man had a side business typing theses; presumably some of the odd alternates are custom dictionary entries related to these papers. Julie Rousseau gives the example of "Pennyhough," which was offered as an alternative for a misspelling of "Trentham." (Pennyhough happens to be the surname of a cleaning lady who reported having seen Lady Prudentia Trentham's ghost.) A web search revealed several authors of scholarly works who are named Pennyhough, so it isn't hard to imagine that a student might have cited the name in a thesis and that it was added to the custom dictionary years back and since forgotten.

Julie Rousseau said that the researchers told her they find some of her explanations far-fetched and do not consider the case closed. It is interesting to come across people who feel that a ghost communicating via a spell-checker is less far-fetched than a software glitch. Nonetheless, kudos to the pair for having IT professionals look into it.

There is a lesson here for both sides of the spirit divide, and that is that hasty assumptions serve no one. To make up one's mind based on nothing beyond a simple summary of events—as believers and skeptics alike tend to do—does nothing to forward the pursuit of solid answers.

DAVE IS STILL talking. I've learned a lot of new things this morning, many of which make me want to raise my hand and go, *What???* just like that, with three question marks. I want to say, *Where's your proof, Dave? How do you know ghosts drain power off your car batteries to manifest themselves? Where did you read that cemetery planners chose their locations to be near "portal openings"? Portal openings!? I'll show you portal openings . . .* But I seem to be the only one having these thoughts, so I keep quiet.

The last topic of the morning is spirit photography. Dave is talking about "spirit orbs," supposed blobs of energy that are very similar to what shows up on your prints if a dust speck or raindrop was right in front of the lens. People often e-mail Dave photos of "spirit orbs" that are obviously dust, and he then has the unpleasant task of telling them this.

"People used to say that when old buildings were renovated, the renovation disturbed ghosts, because they'd get a lot of orbs on their pictures." Dave smiles. "Well, renovation also disturbs dust."

In other words, consider all angles. Take the link between electromagnetic fields and spirits. I called the manufacturer of the TriField Natural EM Meter and asked what made the company believe they had grounds to market it as a Ghost Detector. No studies support this; as Dean Radin puts it, "there is no evidence that EMF meters detect anything other than EMF." The man on the phone said he didn't know for sure, but he assumed that someone, possibly more than one person, had noticed that an EMF meter registered a jump when he or she was standing in a spot that felt haunted.

But what if we take the ghost out of the equation? What if exposing someone's brain to certain types of electromagnetic fields could create a feeling of an invisible presence? What if the energy *is* the ghost?

Aiming to find out, I took my brain to Canada.

Inside the Haunt Box

Can electromagnetic fields make you hallucinate?

IT IS HARD to imagine being terrified in Sudbury, Ontario. A bland, friendly mining city 150 miles north of Toronto, Sudbury is best known for the Big Nickel, a thirteen-ton statue of Canadian pocket change. Curling is popular here. Under the category "Other Fun Stuff" on the Sudbury website, the tourism people have listed a vegetable store.

Nevertheless, fear is on the agenda tonight. I'm heading to the Consciousness Research Lab at Laurentian University to be electromagnetically "haunted." Michael Persinger, the neuroscience professor who runs the lab, has a theory about ghosts. The theory holds that certain patterns of electromagnetic field activity—both the earth's own natural kind and the man-made kind created by wiring and appliances and power lines—can render the brain more prone to hallucinations. In particular, the sort that involve an invisible, sensed presence.

In a study published in 1988, Persinger compared thirty-seven years of dated *Fate* magazine haunting reports with geomagnetic activity for those dates. He found a nice correlation, and he wrote up his findings in *Neuroscience Letters*. In a similar study three years later, University of Iowa psychologists Walter and Steffani Randall examined monthly fluctuations in solar winds (which influence the earth's geomagnetics) to see if they mirrored monthly ups and downs in "humanoid hallucinations" culled from old Society for Psychical Research records. Indeed, both showed peaks in April and September, with a trough in between.

Persinger then turned his attention to man-made electromagnetic fields (EMFs). In 1996, a Sudbury couple had contacted him about strange goings-on in their house. They heard breathing and whispering sounds and at one point felt someone touching their feet as they lay in bed. The husband saw an apparition of a woman who appeared to move through the couple's bed. Persinger and two colleagues drove out to the house and set up equipment to monitor EMFs in the various rooms. True to his theory, the house was an electromagnetic free-for-all. Wires were poorly grounded and circuits overloaded with electronic equipment. Not only were the EMFs most intense in the places where the couple had experienced their "ghosts," but they showed the telltale irregularities that Persinger has come to see as the hallmarks of haunt-prompting fields.

If electromagnetic fields like these could be generating "hauntings," then it's reasonable to assume you could create what Persinger calls a synthetic ghost by exposing people to similar, laboratory-generated EMFs. This is what Dr. Persinger will, at my request, be doing to me tonight. In the back of his lab is a soundproof chamber—a haunt box—outfitted with a comfy chair, in which subjects sit while Persinger directs complex patterns of EMFs into their brains via an electromagnet-bedecked, wire-sprouting helmet.

Normally I object to strangers beaming force fields into my brain. But Michael Persinger has a university post and his papers are published in mainstream medical journals. How dangerous could it be? I've been pondering this in the back of the cab on my way to the university. The driver, a graduate of Laurentian, asks what sort of work I'm doing there. I tell him I'm visiting Michael Persinger. He swivels to look at me. "Strange guy. I heard he's totally nocturnal, eh? And he keeps rats."

It would seem that in Sudbury, Ontario, Michael Persinger is more famous than the Big Nickel. He is assuredly more interesting. "They say he mows his lawn in a three-piece suit."

Dr. Persinger is not at the lab to greet me. I'm told he'll be arriving later, after I've filled out some personality inventories, part of the protocol for whichever experiment my data will become part of. A research assistant seats me at a table just inside the door, and there I remain, for the next half hour, checking True and False boxes. *I have been taken aboard a space ship. I sometimes tease animals. I certainly feel useless at times. There is something wrong with my mind.*

I can't imagine the sort of personality that would check True for some of these statements. Or maybe I can. *I mow the lawn in a suit. I like rats.*

The lab assistant gets up to run an errand. For some reason, she locks the door as she leaves. I sit awhile. I get up and pace. A table against one wall is stacked with plastic storage containers, each holding two large chunks of brain preserved in a clear liquid. *Fascinating*, I think, and then I notice the labels on the lids: "Sean and Kristy." "Michelle and Holly." "Brent and Derek." On any other evening, I'd have assumed that these were the students who dissected the brains. On any other evening, it would not cross my mind that Sean and Kristy might *BE THE BRAINS IN THE CONTAINER.*

The door opens behind me. It's a man in a three-piece suit. The suit is black, with pinstripes and a watch fob. Dr. Persinger, trim and white-haired and seemingly sane, introduces himself. Before he even sits down, he begins to scan my answer sheets. He seems in a hurry to get me into the chamber. I tell him I'd like to ask some questions first. He says that's fine and that we'll walk over to the Colony Room and talk there. I picture dark wood paneling, trustees conversing in hushed tones.

Dr. Persinger unlocks an ordinary-looking door on the other end of the building. Inside are two long walls of cages and a corresponding wall of smell: Colony, as in rats. But the rodents aren't a hobby, as the cabdriver had made it sound; they're subjects in various experiments on the beneficial effects of EMFs. (Persinger believes certain kinds of EMFs can be helpful in treating conditions as widely varying as depression and multiple sclerosis.) He tells me he returns several times during the night to make adjustments and gather data, often going to bed around four. Hence the reputation as a nocturnal being.

While he attends to his rats, Persinger gives me the lowdown on the haunt theory. Why would a certain type of electromagnetic field make one hear things or sense a presence? What's the mechanism? The answer hinges on the fact that exposure to electromagnetic fields lowers melatonin levels. Melatonin, he explains, is an anti-convulsive; if you have less of it in your system, your brain—in particular, your right temporal lobe—will be more prone to tiny epileptic-esque microseizures and the subtle hallucinations these seizures can cause. Persinger adds that the emotions of bereavement produce stress hormones that may serve to raise the likelihood of these microseizures even further.

Persinger isn't the only researcher to have examined the link between spirituality and tiny seizures in the temporal lobe. In a 2002 *Psychological Reports* study of 242 undergraduate volunteers, scores on a ninety-eight-item spirituality assessment were significantly predictive of scores on a questionnaire that assessed a cluster of symptoms of complex partial epilepsy, including hallucinations, fear, and a sense of detachment from one's body. Also known as temporal lobe epilepsy, this condition often goes undiagnosed. Without a medical explanation for these mystifying experiences, patients may interpret them as spiritual events and adjust their belief systems accordingly.

It would seem that Persinger has the whole ghost business neatly sewn up. It's true that people with naturally occurring microseizures—e.g., sufferers of complex partial epilepsy—often have hallucinations. It's also true that EMF exposure dampens the body's natural production of melatonin. This has been shown in rats and in dairy cows housed inside a "bovine exposure chamber" at the McGill University Dairy Cattle Complex* in Quebec.

The results of Persinger's lab work suggest that you can indeed evoke that haunted feeling in a lab using EMFs. Of the approximately one thousand people who have had Persinger's

---

*When I got home, I wrote to the Dairy Cattle Complex researcher, Javier Burchard, and asked him if the cows ever behaved as though there were an invisible presence in the chamber. He replied that he'd never seen any behavior that was so abnormal as to cause him "to pursue research in that direction." This suggested to me that he had in fact seen cows behaving in a *mildly* abnormal manner, so I wrote back again and encouraged him to pass along any anecdotes. I clearly sounded like I had a dairy cattle complex of my own. "I'm sorry," came the exasperated reply. "But I cannot give you any cow story."

signature electromagnetic bursts applied to their right temporal lobes, eighty percent, he says, have felt a presence. In 2002, he published a paper on lab-generated hauntings in the *Journal of Nervous and Mental Disease*. Forty-eight university students were exposed to complex one microTesla electromagnetic fields either over the left temporal lobe or the right, or both. A fourth group received sham pulsations. Those whose right brains were exposed were more likely to report feeling fear and sensing a presence than were left hemisphere or sham exposures. Disappointingly, no other researchers have replicated Persinger's work.

THE SOUNDPROOF chamber where Persinger haunts his guests is about as big as a freight elevator. It appears to have been decorated circa 1970; the floor is covered with yellow and brown shag carpeting, and the subject chair is padded with a messy drape of cheap Mexican blankets. I half expect someone to pass me a bong.

Linda St-Pierre, the lab manager, is sticking EEG leads to my scalp. While she does this, I ask Persinger about a statement he made in one of his papers: "Although these results suggest that these apparitions are an artifact of an extreme state-dependence, the possibility that they are associated with transient, altered thresholds in the ability to detect normally indiscriminable stimuli cannot be excluded." Could the "normally indiscriminable stimuli" he's speaking of be generated by someone dead? In other words, is it possible that—rather than prompting hallucinations—certain electromagnetic field patterns enhance people's ability to sense some sort of genuine paranormal impulse or entity?

Persinger acknowledges that both explanations are possible. It could be that people are being physically affected by the electromagnetic fields and then applying their own cultural overlay ("Ghost!") to explain the experience, and it could also be that people—at least some of them—are suddenly, as a result of the field's effect on their brain, able to pick up, as Persinger puts it, "actual information that's in the environment." Persinger thinks the latter is likely. "Particularly," he says, "in places where people experience the same thing again and again." Before I arrived, I had thought Persinger was a skeptic, a debunker, but clearly he's not.

As he talks, Persinger ties a worn paisley scarf around my head to secure the EEG leads. He says he's been using it for years. He pauses. "Wouldn't it be funny if it turned out that all along, it was the scarf?" Persinger slides an orange Skidoo helmet down over the scarf. Glued to the outside of the helmet are eight small electromagnets, which will deliver the milliseconds-long pulses to my brain. Persinger assures me that the exposure level is no higher than that of a hair dryer. It's the pattern of the signal—its complexity—that matters. Then he shuts off the lights and backs out the doorway.

"Ready?"

No. "Yes."

The door shuts with a heavy, whispering *clumpf*, like a space-station air lock. Five minutes pass. I want to feel a presence, but mostly I just feel absence: of sound, of light, of the eerie effects I'd hoped for. If you've ever waited quietly in the dark before a surprise party begins, then you know what I'm experiencing right now. I worry that I'm going to disappoint Dr. Persinger, much as I'm disappointing myself. *I certainly feel useless at times. There is something wrong with my mind.*

Toward the end of the session, I begin to see and hear some things. Glimpses of faces, utterances that flash through my conscious mind so quickly I can't remember them a second later. At one point, I hear a police car off in the distance—the repeated whoop-whoops that signal a driver to pull over.

After it's over, Dr. Persinger comes into the chamber and sits down on an ottoman to interview me about my experiences. I interrupt him. "Did you hear the police siren?"

"No."

"I did. From over in that direction."

Persinger looks up from his notepad. "This is a soundproof room."

Ah. Then I must have been drifting off to sleep.

"You're labeling," says Dr. Persinger. "Don't label." He gets up to retrieve my EEG printout. He flips through page after page of taut, insistent scratchings. "You weren't drifting off to sleep. Not even close."

Whatever it was and as real as it seemed, it wasn't something I'd interpret as a paranormal phenomenon. About five years ago, for a period of several months, I would occasionally be awakened in the night by someone knocking loudly and insistently at either the front or the back door. So clear and so convincing were the knocks that the first time it happened I got out of bed, put on my bathrobe, and stumbled to the door, greatly amusing my husband, Ed, who was in the living room reading. No one was there. There had been no knocks. That had seemed spookier than this, but maybe it's the context: sitting expectantly in a lab versus lying alone in bed late at night.

Persinger says that based on my answers to the questionnaires, I'm left-hemispherically dominant. I'm a "least responder." For comparison, he hands me a sheet of paper with passages from transcripts of the sessions of highly responsive, right-hemispherically dominant types:

"I felt a presence behind me and then along my left side. . . ."

"I began to feel the presence of people, but I could not see them; they were along my sides. They were colourless, grey-looking people. I know I was in the chamber but it was very real."

Most impressive, to me at least, was the response of paranormal researcher-turned-skeptic Susan Blackmore, who visited Persinger well into her skeptical years, for a *New Scientist* article: "I felt something get hold of my leg and pull it, distort it, and drag it up the wall."

It's possible that the reason I've never experienced a ghostly presence is that my temporal lobes aren't wired for it. It could well be that the main difference between skeptics (Susan Blackmore notwithstanding) and believers is the neural structure they were born with. But the question still remains: Are these people whose EMF-influenced brains alert them to "presences" picking up something real that the rest of us can't pick up, or are they hallucinating? Here again, we must end with the Big Shrug, a statue of which is being erected on the lawn outside my office.

10

Listening to Casper

A psychoacoustics expert sets up camp in England's haunted spots

**IR FULKE GREVILLE** lived in Warwick Castle from 1605 until 1628, the year he was stabbed by his disgruntled manservant Ralph.* The murder happened in London, but Sir Fulke's ghost, in the manner of certain lost but persevering pets, found its way back to the castle. I'm guessing it got a lift from the Tussaud Group,

*A gripping moment that capped an otherwise drab existence. A proponent of what *Encyclopædia Britannica* calls "a plain style of writing," Greville failed to publish much of anything while alive. Well-born but repeatedly passed over for appointments, he was eventually dubbed Knight of the Bath. (The Knights of the Bath are an official Order of Her Majesty the Queen, who does not take enjoyment from Monty Python–style send-ups thereof. Or possibly she does: John Cleese was offered—and declined—an Order of the British Empire.)

the wax museum people, who bought the castle in 1978 and installed a Fulke-inspired production number called "Warwick Ghosts—Alive!" ("unsuitable for anyone of a nervous disposition").

Sir Fulke is Coventry's most famous ghost and certainly its highest-grossing, but not, to my mind, its most intriguing. For that you must visit the home of chartered engineer and psychoacoustics researcher Vic Tandy. Tandy, who teaches at Coventry University, is a big middle-aged guy with a goofball grin and glasses with heavy lenses that tend to pull the frames slightly off-kilter on one side or the other. He fits my stereotype of an engineer so well that when I hear him say things like "I have a second-level in aikido" and "I'm also a magician," I have to stop myself from going, *Really?* We are sitting in Tandy's living room with his wife Lynne and their son Paul, who has a stall at the local market selling rubber dog doo—and probably, knowing this family, a Ph.D. and a Heisman Trophy.

Tandy's ghost story takes place twenty-some years ago, at a nearby factory that manufactured life-support systems. Tandy designed the company's products and he put in a lot of overtime. One night as he returned to his lab from a coffee break, the cleaner barreled past him with a stricken look. "She told me there was a ghost in there. She said she'd been feeling uneasy, as though someone was in there with her, and then this gray thing appeared in the corner of her eye, and she took off running."

Tandy's first guess was that an anesthetic bottle was leaking, and the fumes were causing the cleaner to hallucinate. Everything checked out fine, so he put it down to, as they say at Warwick Castle, a nervous disposition. The next night, though, working late again, Tandy began to feel strange himself. "I felt my hackles go up."* Again, Tandy suspected fumes. He wondered whether someone had left the stopper off the tricoethylene, which his lab mates used for degreasing machine parts. "That wasn't it, so I thought, *Right. I'll go have a coffee*. I came back in. Same thing." Again with the hackles. "Then I see this gray object around to the side of me. I would say I was projecting a form onto it, trying to make sense of it, but . . . it had arms and legs at one point. I turned to look at it and it disappeared. The following day I was going in for a fencing competition—"

"*Really?*"

"Yeah. So anyway, I'd brought my foil in to fix it. I'd put it in the vise and gone over to my desk, and when I turned back, it was moving on its own. Aaaaaa!" I don't know how to spell the sound Tandy just made. Imagine an opera singer being garroted at the crescendo of an aria. What he means is, it scared the bejeebers out of him. But only for a moment. "I thought, *No, no, come on, there* has *to be a reason for this*."

Ever the engineer, Tandy set out to find the answer. Had the cleaner come in at this juncture, she would not have been reassured about the safety and normalcy of her place of employment. She would have found Vic Tandy on his hands and knees, sliding a fencing weapon slowly across the linoleum. Every few seconds, he'd stop to jot notes on a pad. By watching for where the blade started to vibrate, he could

---

*Tandy is speaking metaphorically. Humans don't have erectile hair or feathers on the backs of their necks. Looking into this, I learned that hackle feathers are popular for fly-tying. It took a while to figure this out, because the Google entries would say things like, "This is a Metz Grizzly Hen Neck hackle. It could be used for a Matuka-style streamer wing, however, and it's a top choice for streamer collars, as it's soft and pulses when the barbules are 'unzipped.' "

measure the peaks and troughs of the sound wave he suspected might be his ghost and pinpoint the frequency. (When sound pressure waves hit an object, they cause it to vibrate; if the object is an eardrum, the brain reads these vibrations—a certain range of them, anyway—as sound.)

Tandy's suspicion was that his ghost was the product of inaudible, low-frequency sound waves—infrasound. Indeed, when he set up his measuring equipment in the lab, he found a sharp peak at nineteen hertz. (Infrasound runs from zero to twenty hertz.) If the source is powerful enough, infrasound can, in addition to setting fencing foils aquiver, engender all manner of mysterious-seeming phenomena. Unbeknownst to audience members, infrasound pulses were sent out at certain points during a piano concert at Liverpool's Metropolitan Cathedral in September 2002. It was at these points that concertgoers reported—via a questionnaire distributed before the concert began—a variety of physical effects, such as tingling on the back of the neck and "strange feelings in the stomach," as well as an intensification of their emotions.

Infrasound has also been reported to cause vision irregularities: sometimes blurring, sometimes a vibrating visual field. The eyeball, Tandy explains to me, has a resonant frequency of nineteen hertz. Meaning that in the presence of a standing nineteen-hertz infrasound wave, your eye would start to vibrate along with the waves. This is similar to the effect of a powerful operatic voice on a wineglass: When the voice—created by pressure waves from vibrating vocal cords—hits on the resonant frequency of glass, the glass in turn begins to vibrate and may, if the note is held, even shatter.

Tandy explains that peripheral vision is extremely sensitive to movement, a helpful adaptation for dealing with predators that sneak up on you from the side. "If your eyeball is dithering, the sides—the peripheral vision—are where it's going to register." The blurry gray ghost in the edge of the cleaner's vision could have resulted from just such a dither.

Next Tandy went off in pursuit of the source. He found it in the basement. "The maintenance people had replaced an exhaust fan," he says. "I think they made it themselves. Huge, huge amount of unneeded energy. I mean, it was quite surprising the fan wasn't standing still and the building going 'round it."

All of this got Vic Tandy thinking. What if he were to visit some of England's purportedly haunted spots and take some sound readings? What if the feelings people report when they think they've been in the presence of spirits are in fact the effects of infrasound? The more he thought about it, the more sense it made. Old buildings have thicker, more solid walls, which resonate better. And old abandoned castles and cellars often have no furniture or curtains to absorb sound waves. Infrasound would also help explain why reports of ghosts are often localized—why people sense a presence in just one part of a room. Infrasound tends to "pool"—it registers strongly in the spots where the peaks and troughs of sound waves overlap, and disappears where peak and trough cancel each other out. Tandy even has an infrasound-based explanation for why people sometimes feel cold in the presence of what they take to be a ghost. Infrasound can activate the fight-or-flight response, and part of that response is a curtailing of blood to the extremities. Hence the chills (and the racing heart and thus, it stands to reason, the unease).

Tandy knew from local hearsay that the Coventry Tourist Information Centre was a promising place to start. Though they spent their days pushing "Warwick Ghosts—Alive!" the staff was convinced that something ghostly was going on directly below them. Excavations for the foundation of the tourism office had uncovered a fourteenth-century cellar that the tourism staff now uses for storage. In a *Journal of the Society for Psychical Research* article on the project, Tandy quotes a Coventry tour guide who had accompanied a Canadian journalist to the cellar: "The gentleman was frozen to the spot and the colour drained from his face."

Tandy went in to take some measurements. I asked him if he felt anything. He said only once—a brief, sudden sense of something "washing over him." His wife Lynne, who has accompanied him on several visits, volunteers that she felt nothing. "Though I do sometimes feel a strange oppressiveness in the Sainsbury's dairy area." I volunteer that, owing to her accent or my fourth-grade maturity level, this came through as: "the Sainsbury's derriere." Lynne's look suggests that the humor isn't registering. It suggests she might think I'm something of a dairy area myself.

Tandy did not find infrasonic frequencies in the cellar, but he did find them just outside it. The eighteen-hertz entity lives in the hallway that opens into the cellar (though its source remains unknown). Tandy figures people were blaming the cellar because of how it looked. As he puts it, "You don't get ghosts in well-lit white-walled concrete corridors. You get ghosts in vaulted fourteenth-century cellars."

According to a Dortmund University nonlethal weapons expert named Jürgen Altmann, infrasound can, in a small percentage of the population, set off vibrations in the liquid inside the cochlea. These vibrations—which happen because of an uncommon anatomical weakness in the bone structure of the ear—could create a sudden, inexplicable feeling of motion, which could lead to the unease that some of the cellar visitors reported.

The majority of visitors, however, feel nothing. Tandy

gave a talk earlier in the week and took the entire crowd over to the cellar afterward. Despite their having been primed to feel something, only one out of the group of fifty did. The same meager odds appear to apply for industrial infrasound. NASA astronauts on liftoffs are exposed to massive infrasound vibrations, to no apparent deleterious effect. (It was, in fact, in a NASA contractor's report that Tandy read about the vibrating eyeball effect. NASA had experimentally exposed volunteers to infrasound back in the sixties, to be sure, as Tandy puts it, "that they didn't deliver jam to the moon.") It is thought that only a small portion of the population is sensitive to infrasound. Tandy believes that when the odd office worker starts talking about Sick Building Syndrome, infrasound may in fact be the culprit. There are said to be people so debilitatingly sensitive to infrasound that even the very low levels of it that come off the ocean can make them nauseous. At any rate, it's not the sort of situation where you can set up a speaker and inflict mental and physical discomfort on demand.

This particular fact came as a great disappointment to the military-industrial complex. For years, infrasound was served up as the next big thing in nonlethal weapons. Obviously, powerful amplifiers would be needed to boost the decibel level—unless your intent was simply to make your enemy feel peculiar. In strong doses, infrasound has been alleged to cause all manner of bodily unpleasantness: nausea, salivation, "extreme annoyance," rapid pulse, vibrating visual field, "intolerable sensations in the chest," gagging, vomiting, bowel spasms, and "uncontrollable defecation." Jürgen Altmann, the best authority on the subject that I could find, says that the more dire second half of the list is hearsay. In the vast stack of literature that Altmann reviewed, he found only one allegation of vomiting and none for bowel spasms and their pal uncontrollable D.

Contrary to persistent Internet rumors, actual infrasound weapons are rare. Altmann found one Russian institution—the very specifically named Center for the Testing of Devices with Non-Lethal Effects on Humans—that was said to have developed a device propelling a baseball-sized pulse of about ten hertz over hundreds of meters. He could find no information on the efficacy of the device, but his tone suggests you'd be better off propelling real baseballs.

Nonetheless, people worry. "I still get people ringing me up, thinking their neighbor is trying to get them out of the house by shooting infrasound at them," says Tandy. I used to have a neighbor who shoots high-decibel Eagles songs out his windows, causing nausea and extreme annoyance at a fraction of the cost. I'd have loved to get my hands on a retaliatory infrasound blaster.

"Try a church organ," suggests Tandy. "The big ones put out a lot of infrasound. Or you could rent an elephant."

Elephants, as well as whales and rhinos, have recently been found to communicate by infrasound; they can both produce it and hear it. In the wake of recent research at the Fauna Communications Research Institute in North Carolina, tigers were added to the list. Tigers have large territories to defend, and it's thought they use infrasound—which has the advantage of carrying over long distances and penetrating dense foliage—to warn trespassers.

The tiger finding spoke to Tandy. The fact that humans—albeit a minority of them—are able to sense infrasound had puzzled him. Why would the ability have evolved if we don't communicate in infrasound? Perhaps to sense predators. Being able to detect a tiger in the vicinity was—for primitive man, anyway—a valuable knack. "So maybe there are people in the cellar whose tiger detector, as it were, is going off."

The research is a nice fit with Tandy's work. Though tigers' vocalizations were found to span a range of audible and inaudible (to us) sound, their roars were measured at and below eighteen hertz—very close to the infrasound frequency that set Vic Tandy's saber rattling. To test the notion that tigers use infrasound to ward off potential rivals, Elizabeth von Muggenthaler, the researcher who had recorded vocalizing tigers, set up powerful speakers to play the big cats' roars and growls back to them. If ever there were a moment when acoustical science broke through the drear confines of audiograms and spectral analyses, this was it. Von Muggenthaler reports that the recordings caused several of the tigers to "roar and leap towards the speakers."

You can try this experiment on yourself by turning your computer speaker to top volume, going to www.acoustics.org/press/145th/Walsh2.htm, scrolling down to the paragraph about roars, and clicking on the speaker icon. Even though I know what's coming, it scares the rubber dog doo out of me every time I play it. I recall once going to the big cat house of our local zoo at feeding time. For a solid minute, the tigers and lions stood in their cages and roared. I started to cry, though I wasn't upset. I had the same embarrassing response to what I'm guessing were the effects of infrasound twice before: once while standing on a rooftop that was buzzed by a Blue Angels fighter plane, and another time standing two streets away from an imploded building as it collapsed. Also, I used to feel an ineffable queerness in my chest during Sunday mass, which I put down to God looking inside me and knowing I wasn't listening. Now I'm thinking it was the organ music. I'm thinking I must be an infrasound sensitive.

I'll soon know, because Tandy has promised to expose me to some nineteen-hertz waves. In fact, that's where we're heading now. Tandy gets up from the couch. It's 6:30 p.m., and the Coventry tourism office has closed. Perhaps we're going to the haunted tower at Warwick Castle? Tandy stops halfway down the first-floor hallway and makes a left. We are not going to Warwick Castle. We're going to Vic and Lynne Tandy's dining room. "You can really get a nice standing nineteen-hertz wave going in here. I've got my kit all set up for you."

Tandy's laptop is set up to channel computer-generated infrasound through the subwoofers of a car stereo amplifier and into a speaker. He pulls out a dining room chair for me. The speaker is seated at the head of the table. "Ready?" says Tandy.

He hits a series of keys on the laptop. We sit in the quiet, heads bowed, as if waiting for the speaker to say grace.

I think I feel something, but then again, I'm looking for it. Tandy says he can't usually feel it while it's on, but that he notices the room feels different when he shuts it off. So we turn the infrasound off, then back on, and then off again. It's hard to say. It's certainly subtle.

Lynne comes in to set the table for dinner. After she leaves, I ask Tandy to put the infrasound on one last time. He leans over, presses some keys. Is there a mild buzziness in my brain? A faint, indescribable weirdness? It's there, I think it's there. "I can feel something already," I whisper.

Tandy looks up from his keyboard. "I haven't got anything on yet."

IF YOU ASK me which is the more likely explanation—infrasonics or spirits—I will tell you to apply the wisdom of Occam's razor,* a principle which holds that the simplest, the least far-fetched, of two competing theories is the place to put your money. But depending on who's shaving, Occam's razor yields manifestly different views. To those who believe in an afterlife, the most straightforward explanation for hearing your dead dad is that you're hearing your dead dad's spirit. Infrasonics and vibrating eyeballs and fight-or-flight responses would, given this particular worldview, seem to be needless and unlikely complexities. But to those of a less spiritual bent, the concept of a consciousness leaving a body and persisting in some ordered form that is able to interact with living beings is a notion that demands an even more elaborate and unnecessarily complex explanation.

Perhaps it's time for a break from the tail-chasing complexities of scientific method. Perhaps some other learned pursuit has something to offer us. Has anyone, for instance, tried to prove the existence of ghosts in a court of law? In fact they have. In the farm belt of central North Carolina, some eighty years ago.

---

*The principle known as Occam's razor was not, curiously enough, William of Occam's idea. Occam simply used it—frequently and "sharply," to quote the *Encyclopædia Britannica* entry—so much so that it became known as his razor. The entry goes on to say that "he used it to dispense with relations, which he held to be nothing distinct from their foundation in things; with efficient causality, which he tended to view merely as regular succession," a sentence that cries out for Occam's editing pencil.

Jas L Chaffin (seal)

this January 16 1919

11

Chaffin v. the Dead Guy in the Overcoat

In which the law finds for a ghost, and the author calls in an expert witness

IN THE SUMMER of 1925, the ordinary life of a Mocksville, North Carolina, farmer shifted a few acres shy of ordinary. James Pinkney Chaffin lived with his wife and daughter in a four-room house on a stream in a field that he planted with sugarcane and cotton. Chaffin picked and baled his own cotton and made molasses from the sugarcane he grew. He carried the molasses in jugs on his back to sell to his neighbors and the townspeople in Mocksville. He did the same with the butter his wife made and the axe handles that he carved and sold for twenty-five cents. On Sundays he walked two miles with his family to the Ijames

Baptist Church, where he sat each week in the same seat, beside an open window—"so he could spit his tobacco," recalls his grandson Lester. Evenings, James Pinkney Chaffin sat by the fire and greased his boots and sharpened his blades. He did not drink or smoke. He was, says Lester, "just as plain as an old shoe."

One morning in June 1925, James Pinkney Chaffin announced to his wife that his father—who had been dead four years—had appeared to him at his bedside. Chaffin was not given to dreams of prophecy or to ghost stories or practical jokes, and one can imagine that the breakfast mood that morning was a bit strained. He confided to his wife that several times over the past month, he had dreamed of his father, James L. Chaffin, appearing at his bedside with a sorrowful expression. The previous night, his father had appeared in a black overcoat,* which the son recognized from when his father had been alive. James Senior stepped closer to the bed and opened up one side of his overcoat, in the manner of a man selling purloined watches. "He pointed to the inside pocket," Chaffin is quoted as saying in Mocksville's *Davie Record*, "and he said: 'You will find something about my last will in my overcoat pocket.'"

---

*SPR cofounder Frederick Myers muses at some length upon "the question of the clothes of ghosts—or the ghosts of clothes. . . . If A's phantom wears a black coat, is that because A wore a black coat, or because B [the person who sees A's ghost] was accustomed to see him in one? If A had taken to wearing a brown coat since B saw him in the flesh, would A's phantom wear to B's eyes a black coat or a brown? Or would the dress which A wore at the moment of death dominate, as it were, and supplant phantasmally the costumes of his ordinary days?" Myers's guess is that A triggers a remembered image of himself in B's mind, and that therefore A's ghost would be clad in black, and not the brown coat he wore when B wasn't around, or his funeral suit, or the field hockey kilt C liked him to put on when he'd had one too many glasses of port.

At the time, as far as anyone knew, the last will of James L. Chaffin was the one on record in the Davie County Clerk's Office, dated 1905. In a perplexing act of filial betrayal, the old farmer had directed that his entire estate—farmland amounting to one hundred and two acres—go to his second-youngest son, Marshall. Nothing was left to James Pinkney Chaffin or his elder brother John, or to the youngest of the four sons, Abner. To John, especially, it was an egregious slight, as land in that day was typically bequeathed to the eldest son. Though the three sons must surely have been bitter about the will, they did not contest it.

After some searching, Pink, as he was known to his family and friends, located the old man's overcoat, in the attic of his older brother John. "On examination of the inside pocket," his testimony goes, "I found the lining had been sewed together. I immediately cut the stitches and found a little roll of paper tied with a string which was in my father's handwriting and contained only the following words: 'Read the 27th Chapter of Genesis in my daddie's old Bible.'" (Chapter 27 is a parable of two brothers, one who cheats the other out of his rightful inheritance.) With his daughter Estelle and his neighbor Thom Blackwelder along as witnesses, Pink proceeded to his mother's house, where they found the old Bible in the attic. Blackwelder opened the dilapidated book to Genesis and discovered that the facing pages that make up Chapter 27 had been folded over to embrace a single piece of ruled yellow tablet paper. It was a second will, dated 1919 and dividing the land equally among the four children. Marshall was by now dead—he died from a faulty heart valve less than a year after inheriting his father's land—but his wife Susie, described by grandson Lester as a more "downtown" sort of person than any of the Chaffin brothers, immediately contested the second will. A date for a trial was set.

The story spread—as stories combining ghosts and large chunks of money and feuding relatives will—and by the time the day for the trial arrived, members of the press were thick as the flies in Pink Chaffin's unscreened living room. Pink arrived in court with ten witnesses in tow—family, friends, and neighbors—all prepared to attest that the signature on the second will was indeed that of James L. Chaffin. (The will itself bore no witness signatures.) After the jury was sworn in, the judge called a lunch break. Apparently Susie and the brothers reached a deal during the recess. In a move that stunned and deeply disappointed the gathered crowds of reporters and townsfolk, Susie stated that the signature was genuine and withdrew her opposition. The widow and three brothers had agreed to share the estate equally. The court thus formally decreed that the document in question—a paper whose secret location had been pointed out by an apparition—was indeed the last will and testament of James L. Chaffin.

Though the reporters were denied the gleefully anticipated spectacle of shouting, finger-pointing loved ones, they left with an even better story. "Dead Man Returns in Dream," ran a local headline. "Can the Dead Speak from Grave?" asked another.

About a year later, Britain's Society for Psychical Research got wind of the case and hired a local lawyer to interview the parties involved and submit a report. The lawyer, J. McN. Johnson of Aberdeen, North Carolina, said he held "scant respect" for the beliefs of SPR members, but promised to pursue his task with mind held open. He obtained sworn statements by James Pinkney Chaffin and Thomas Blackwelder, the man who had driven Pink and Pink's daughter in his Model T on the twenty-mile journey to find the old coat and, later, the grandfather's Bible. Johnson was impressed with the sincerity of the Chaffin clan. "I believe I am safe in asserting that if you once talked with these honest people and looked into their clear, unsophisticated countenances, your criticism would vanish into thin air, as did mine." He wrote these words in a 1927 letter to the SPR, and concluded that the will was genuine and the farmer's ghost story improbable but true.

Johnson ruled out the possibility that the second will was a fake on the grounds that not only the witnesses but the defendant herself, Marshall's widow Susie, agreed that the handwriting on the second will was that of James L. Chaffin.

You would think that the SPR would need no more convincing. You would think that a letter like this, following a courtroom victory, would be trumpeted in the pages of the SPR's journal as proof positive of the soul's survival after death. But you would be wrong. In response to his report, Johnson received a contrary ten-page letter from SPR honorary officer W. H. Salter, which remains to this day in the Chaffin Will file in the SPR archives. Salter felt—and you'd have to agree with him—that the case presented puzzling irregularities. If the old farmer had changed his mind and now wanted his land divided among all four sons, why would he hide the new will so carefully and not tell any of his sons—indeed anyone at all—about it? Wrote Salter: "There is, I admit, no limit to the folly of testators or the secretiveness of farmers, but the present testator seems to have pushed both these characteristics to the limit. But for the apparition, his testamentary wishes would never have been carried out, and one can hardly suppose that during his life he counted on being able to appear as a ghost."

The SPR party line on apparitions is outlined in SPR cofounder Frederick Myers's seven-hundred-page opus *Phantasms of the Living* (which includes a chapter on phantasms of the dead). Myers felt that most are the products of the viewer's own mind. Especially suspect is "an apparition which seems to impart any verbal message," as did the ghost of James L. Chaffin; these are described as "very rare."

Attorney Johnson replied to Salter's letter with a possible explanation. Johnson had been told by a neighbor that the old farmer lived "in mortal terror" of his daughter-in-law Susie, who had in her possession the 1905 will. Changing his will would have meant confronting her, a task James L. was no doubt loath to undertake. So perhaps he hid the new will and planned to tell his three sons about it in his dying moments, so that in death he might escape the wrath of Susie. And then, I suppose, he misjudged his timing, and died before he could tell them. "This man J. P. Chaffin is an honest man and he thoroughly believes his father's spirit appeared to him and gave him the clue to the 1919 will," concludes Johnson's letter. "And his manner appeared to me to be entitled to such respect that to doubt him would be to sin against light."

Salter didn't buy it. He came up with his own scenario, which held the will to be a fake, yet salvaged the innocence of James Pinkney Chaffin. He imagined that the eldest son, John Chaffin, perhaps with the help of his brother Abner, faked the will and the slip of paper in the overcoat pocket. James Pinkney Chaffin was made the unwitting pawn in the plot, for it was he who would be moved to discover the will. This would be accomplished by making Pink believe he'd seen his father's ghost, when what he'd really seen was *his brother John dressed up in his father's overcoat.*

And there the mystery lay. Until April 2004, when yours truly decided to take a trip to Mocksville. I would talk to the descendants of the Chaffin brothers and unearth the two wills. I would hire a forensic document examiner, the best in the business. I would let science decide, once and for all, if the second will was a forgery and our overcoat-wearing ghost a fabrication.

THE DIRT ROAD along which James Pinkney Chaffin walked with his molasses and his axe handles is now four lanes wide. Yadkinville Road grew up to be the shopping mall strip, the predictable, just-outside-town plop-down of Burger Kings and BoJangles. My room in the Mocksville Comfort Inn looks out onto this road, and I try to picture old Pink shambling along with his load, vest fronts flapping in the after-blast of passing four-ton Chevys.

There are fewer farms in Mocksville today, and no farmers at all in this branch of the Chaffin clan. Pink's grandson Lester Blackwelder is a retired Ralston-Purina salesman. He has a salesman's smile, accessorized with a wink and a toothpick. His sincere, clap-you-on-the-shoulder congeniality served him well in his career; he and his wife Ruby Jean live comfortably in a roomy house on an upscale street. Pink's grandson Lloyd is an engineer with Ingersoll-Rand. Neither man so much as grows lemons in the backyard. This all came as a surprise to me, having spoken to Lester and Ruby Jean by phone and having placed them—mostly because of their accents and the "might-coulds" in their speech—in homey farm kitchens with gingham curtains and eggs in wire baskets on the counter.

This afternoon, Lester and Ruby Jean and I have gone visiting. We're sitting in Lloyd Blackwelder's living room, and the two men are reminiscing. (Lester and Lloyd are James L. Chaffin's oldest living descendants; Marshall and Abner have no living descendants, and John's living descendants are the next generation down—too young to recall any details.) Lester was a teenager when Grandpa Pink used to tell him the story of the dream and the will. His mother Estelle rode in the Model T with her daddy Pink the twenty miles to John's house, to look for the overcoat. "Dirt roads the whole way," Lester is saying. "No windows on the car. Mama said she remembered what the coat looked like. The pocket was hand-sewed and there was dirt dobber nestin' all over it."

I don't know what this means and apparently they can tell, because Ruby Jean sets down her iced tea and says, "Wasps' nests, Mary." Sometimes it's just the accent that loses me. "Pie safe" required four repetitions and a trip to the kitchen.

I ask them what James Pinkney Chaffin looked like. "He was real thin," says Lester. "Six foot. Rugged. Mustache. Not a good-looking man."

Ruby Jean twirls the ice in her glass. "He didn't have no mustache, hon."

Lester considers this. He juts his lower lip. "I thought he had a mustache."

Later, Ruby Jean finds a photograph for me: Pink Chaffin and his wife and baby daughter posing in their Sunday clothes. Pink's striped shirt looks new and his hair is combed and pomaded, but you can see the dirt worn into the rims of his fingernails. He confronts the camera with a calm, somber, baldly direct gaze, probably the same one that so impressed attorney Johnson. He doesn't have a mustache.

Lloyd is the younger of the two grandsons. He is dressed in Levi's and a corn-colored polo shirt. His memories of Pink are a child's memories; he recalls the time he sat in his grandfather's lap, the rocker rocking so hard it flipped over backward, and the toy horses Pink made out of pieces of dried cornstalks, with tufts of cotton for the manes. Lloyd crosses the room to a glass-fronted curio cabinet and takes down a glass walking stick, twisted at the neck and bobbed into a rounded handle. "Here's his Sunday cane," he says. The glass is ribboned with red and blue, like stick candy. Having seen only posed sepia photographs of these people, I find it hard to add this colorful, foppish item to the grimy, monocolor tableaus of James Pinkney Chaffin that I've built in my mind. They might as well have shown me the man's floral nosegay and spats.

Neither Lester nor Lloyd remembers his great-uncle Marshall, the original recipient of James L. Chaffin's entire estate. Lloyd recalls a vague aura of ill will surrounding Marshall's wife Susie. Susie, to refresh your memory, is said to have been the one in possession of the earlier will, the one that left everything to Marshall. Interestingly, the first will was written the year after Marshall married Susie. Perhaps Susie pressured her father-in-law into drafting the will. Lester says Grandpa Pink loved to tell the story of the apparition and the secret will, but he can't recall hearing him say anything at all about Marshall and his wife, or the circumstances of the first will. The one thing he recalls is that James L. Chaffin lived with his son Marshall after his own house burned down, so perhaps the father felt beholden to his son. Indeed, Marshall is listed on the "informant" line of James L. Chaffin's death certificate, which suggests a closeness between the two.

Lloyd and Lester are not open to considering that Grandpa Pink might have made up the dream and been part of a plot with his brothers to forge a new will and take back the land. That the ghost and the overcoat and the Bible were all just elements of an imaginative scam dreamed up by the three spurned brothers.

"Pink would just never have thought of that," says Lester.

"Nope," says Lloyd. "He would have considered that crooked."

The old farmhouse where Pink lived when he had the

visions of his father still stands, and Lloyd and Lester offer to take me there. Lester and Ruby Jean squeeze into my rented Hyundai, and Lloyd and his son Brad follow in Lloyd's truck. Lester is driving the Hyundai, so that I can take notes while we talk. At one point he puts the left turn signal on, though there's no road or driveway in view on our left, just an open field of tall grasses. The house sits on the far perimeter of the field, and that is where Lester is headed. "Used to be some tracks here, but not no more." The weeds brush the underside of the Hyundai, making worrisome car-wash sounds inside the car. Lester and Ruby Jean seem accustomed to driving in fields. "Lester, there's the old persimmon tree," trills Ruby Jean.

"Uh-huh," says Lester. He drives in overgrown fields at more or less the same speed as he drives on asphalt. "Grandma made the best persimmon pie, didn't she?"

One side of Pink's house is obscured by a thick climb of honeysuckle. Parts of the house are down to framing now, partly because it's been abandoned so long, and partly because Lloyd pulled some boards away to make a pie safe. The men take me on a tour, pointing out the kitchen, their mama's courting room, the bedroom, the outhouse, the earthen wells to keep the milk cool. There's a doorway out the back wall of the bedroom. If John A. Chaffin came out here to play ghost in his father's overcoat, that's probably how he'd have come in. I tell Lloyd and Lester about the SPR officer's theory. "Har," says Lester. "I doubt that. John was just like Pink. Didn't talk much. Didn't go for foolishness."

IS IT POSSIBLE to dress up like a ghost and fool people into thinking they've seen the real deal? Happily, there is published research to answer this question, research carried out at no lesser institution than Cambridge University. For six nights in the summer of 1959, members of the Cambridge University Society for Research in Parapsychology took turns dressing up in a white muslin sheet and walking around in a well-traversed field behind the King's College campus. Occasionally they would raise their arms, as ghosts will do. Other members of the team hid in bushes to observe the reactions of passersby. Although some eighty people were judged to have been in a position to see the figure, not one reacted or even gave it a second glance. The researchers found this surprising, especially given that the small herd of cows that grazed the field did, unlike the pedestrians, show considerable interest,* such that two or three at a time would follow along behind the "ghost." To my acute disappointment, "An Experiment in Apparitional Observation and Findings," published in the September 1959 *Journal of the Society for Psychical Research*, includes no photographs.

Several months later, the researchers revised their experiment, changing the venue and adding "low moans" and, on one occasion, phosphorescent paint. One trial was set in a graveyard right off a main road and clearly in the sight line of drivers in both directions. Here observers hid in the bushes not only to record reactions, but to "avert traffic accidents" and "reassure anyone who became hysterical." But again, not a single person of the hundred-plus who saw the figure thought it

---

*This comes as no surprise to yours truly, who has twice, on separate continents, carried out an experiment designed to prove the considerable curiosity of cows. This is an experiment I urge you to repeat, simply for the giddy thrill of it. Go into a pasture where cows are grazing in the distance. Shout to get their attention, and then suddenly lie down. The moment you do, they will hurry over to investigate, encircling you and staring down at you with unmitigated bovine fascination.

was a ghost, including two students from India. "Although we are superstitious in our country," the men told one of the researchers, "we could see his legs and feet and knew it was a man dressed up in some white garment."

In their final effort, the research team abandoned traditional ghost-appropriate settings and moved the experiment into a movie theater that was screening an X-rated film. The author of the paper, A. D. Cornell, explained that the X rating was chosen to ensure no children were traumatized by the ghost, as though that somehow explained the choice of a porn theater as a setting for a ghost experiment. This time the "ghost" walked slowly across the screen during a trailer. The phosphorescence was not used this time, and presumably low moans were deemed redundant. No mention is made of the specific images showing on the screen behind the ghost, but clearly they were a good deal more interesting: The audience was polled after the film, and forty-six percent of them didn't notice the man in the sheet. Among those who did, not one thought he'd seen a ghost. (One man said he'd seen a polar bear.)

And so we can safely conclude that if John Chaffin had attempted something as uncharacteristic as dressing up as his father's ghost and moaning in his brother's bedroom doorway, James Pinkney Chaffin would not have been convinced. Though his cows, were they in a position to observe, would have been fascinated.

I HAVEN'T SPENT much time in the South, and I didn't realize how helpful people are there. They help you even if you don't ask for help. I went to Food Lion yesterday, and the checkout clerk told me my yogurt was on special if I had an

MVP card. "Trudy," he said to the bagger when he found out I didn't. "Give me your MVP card." It's the kind of place where you call a total stranger on the phone, and his wife will say, "Hang on, I'll go run and see if I can catch him before he goes off on his tractor!" The closest thing to impoliteness that I've come across so far has been a license plate holder ordering me to EAT BEEF. Eat beef, *please*, I chide.

Thanks to southern hospitality and the kindness of strangers, finding the Chaffin wills turned out to be as simple as telephoning the records office. The woman who answered put me through to the clerk of the court, who regularly picks up his phone. The clerk, Ken Boger, said the old records were in the courthouse basement and I could come down any day of the week and he'd help me find them.

Today is that day. I'm meeting a Tennessee-based questioned document examiner and forensic handwriting expert named Grant Sperry. I found Sperry through the American Society of Questioned Document Examiners, of which he's president. Sperry has been an expert witness in some three hundred federal and state cases, including the Waco mess, where his testimony resulted in the conviction of an assistant U.S. attorney who had denied any knowledge of the pyrotechnic devices used on the compound. Sperry found imprints of his notes about the devices on the page below the page where he'd written them. (Note to the careless: These guys can read the imprints of your writing on a pad as much as ten pages down from the page you wrote on.) Sperry was coming to North Carolina to visit his parents and the Chaffin case intrigued him, so he agreed to help me out for about, oh, 1/100th of what he charges the trial lawyers.

We're waiting at the metal detector in the front vestibule, along with a pile of Sperry's equipment. We've been here several minutes. After a while, a man in a security uniform spots us. "Ain't been nobody manning that for two months." He waves us in.

It's a busy Monday morning, but Boger gets right up from his desk to take us to the basement. Within five minutes, we've got both wills. Sperry sets up a makeshift desk on a stack of boxes full of old case files. Most are boxes designed to hold files, but one says: ORRELLS WHOLE HOG SAUSAGE. Sperry puts on bright blue latex gloves, picks up the wills one at a time, and lays them on a scanner. Now he'll be able to look at them side by side on his computer screen and line up any two elements he chooses. Since both wills are handwritten, we'd thought that we had two lengthy handwriting samples to compare, but Sperry quickly determines that the body of the first will has been written out by someone other than the signer—presumably a lawyer, for the document is written in standard-issue legalese on legal-sized paper. The second will, he says, is all one handwriting. This one is a curious mixture of legalese and down-home sap, penned on a page from a ruled school tablet:

> After reading the 27 Chapter of genesis I James L Chaffin do make my last will and testament and here it is i want after giving my body a desent burial my little property to be equally devided between my four children if they are living at my death both personal and real estate devided equal if not living give share to there children and if she is living you all must take care of your mamy now this is my last will and testament. Wit my hand and seal on other side
>
> Jas L Chaffin

Sperry can tell right away that whoever wrote out the second will wasn't attempting to copy someone else's handwriting. The writing is too fluid and relaxed, too swiftly and confidently written to be a forgery. Forged writing is more like drawing, he says. The person moves slowly and deliberately, stopping and starting and sometimes even touching up letters. I read about this in *Questioned Documents*, a classic in the field, written by the enormously learned and occasionally crabby Albert Osborn. "A genuine writing does not often suggest that the writer is thinking of what he is doing with his pen," Osborne wrote, "while a dishonest writing, when examined with care, often shows quite conclusively that the writer was thinking of nothing else. . . . This is another subject beyond the understanding of the stupid observer." It's clear the second will was written with a relaxed hand. So either James L. Chaffin wrote the second will, or it was written by someone who wasn't overly concerned with creating a convincing match for Chaffin's hand.

Sperry moves on to a comparison of the James L. Chaffin signatures on the two wills. It's likely that the first will was indeed signed by Chaffin, as there are two witnesses. Sperry's job now is to see if the writer of the second will, which had no witnesses, is also James L. Chaffin. The task is complicated by the fourteen years that span the two documents. Handwriting—especially signatures—often changes over time.

Nonetheless, Sperry has reached a conclusion. "There's an old axiom we have," he says, peeling off his gloves. "You can't write better than your best." In other words, I could never do a convincing forgery of my mother's signature. My mother had beautiful, gliding, even penmanship, and mine has always been rat scrabble. She could forge mine, but not vice versa. Once you reach your "age of graphic maturity"—usually sometime in your teens—you've hit the peak of your ability and are unlikely to get much better. If anything, your writing gets worse: Handwriting deteriorates with old age and its decrepitudes—bad vision, stiff fingers, hand tremors.

In the Chaffin case, the situation is backward—and thus suspicious. The skill level in the signatures on the 1905 will is substantially worse than in the 1919 will. Which doesn't make sense if it's the same writer. Sperry pulls up the 1905 signature, written when Chaffin was in his fifties. The letters are awkwardly formed and there are hesitations—not the type of hesitations that suggest forgery, but the type that suggest this person is not a highly skilled penman. That seems likely, given the state of education in Davie County around that time. According to James W. Wall's *History of Davie County*, illiteracy was common among rural families in the mid-1800s. In 1860, when James L. Chaffin was fifteen, only 690 of 1,230 school-age boys in Davie County were enrolled in public schools, and the school year was just a few months long (in winter, when the fields lay fallow). Lester says Grandpa Pink only went as far as third grade, and had a total of nine months of schooling. It's likely his father would have had even less.

"Now look at the later will," says Sperry. "The letter formations are much more fluid. Look at the *f*s. How much less awkward they are." And here Chaffin would have been seventy. "If the J. L. Chaffin signatures on the 1905 will are representative of that particular writer's skill level, and I see no evidence that they are not, then that writer could not have written the signature on the 1919 will." It would seem to be a fake.

Sperry also finds some of the language of the second will suspiciously sophisticated for a nearly illiterate farmer. "Wit" is legalese, as is the phrase "both personal and real estate divided equal."

Sperry highlights a line in the 1919 will. "Look at the wording here," he says. "He wants his property to be divided between his four children *'if they are living at my death . . . and if not living give share to there children . . .'* Let's say the 1919 will was forged and backdated by Chaffin's other sons in an effort to get the land back from Susie Chaffin after Marshall died. We know there was some ill will between her and the other brothers. With that clause in place, she's technically out of the picture: The will leaves Marshall's share of the land to their son, not to her. So let's imagine the scene on the day of the trial. The family goes to lunch—which is a matter of record—and the brothers sit her down and spell it out: 'We've got ten witnesses prepared to vouch for this signature. You've got your choice, Susie. You can go in there and agree that it's his handwriting and we'll cut you a one-fourth share—even though you're not entitled to it in this new will. Or you can let the jury decide, and risk losing it all.' "

Sperry's theory makes some sense. And if the forger's intent was to corner Susie and force her hand—rather than actually convince her of the second will's authenticity—then the breezy, unconcerned handwriting makes sense. Why bother fooling her, if you've got her where you want her?

WHOEVER CHOSE THE epitaph for the gravestone of James L. Chaffin would seem to have had a twinkle in his eye. It says: THY WILL BE DONE. Lester and Ruby Jean and I are out at the Ijames Baptist Church cemetery, visiting the family plots.

Lester has wandered over to the grave marker of a local acquaintance. "He shot hisself on the back porch." He continues on down the rows, narrating death in the flat, evenly paced tones of a stock report. "There's that baby died in the four-wheeler accident. And Thom's son there: struck by lightning on a combine—" Ruby Jean cuts him off. "Look at this, Mary. Man put his two wives on the same tombstone. Wonder how they'd have felt about that!" Three stones down is the grave of one Flossie Gobble. You don't have to meet some people to know you'd like them, and Flossie Gobble is one of those people.

I tell Lester and Ruby Jean about what the handwriting expert found. I am careful to add that Sperry compared Pink's signature on the court papers with the questioned James L. Chaffin signature, and it isn't a match.

"So Grandpa Pink didn't do it," says Ruby Jean. She sounds relieved. I don't add that he must have played some role in the brotherly ruse—unless we buy the scenario of John or Abner Chaffin dressing up in the overcoat and playing ghost.

"Hunh," says Lester. "Do you feel good about it?"

I tell him I'm not surprised by Sperry's conclusion about the signatures, but I am disappointed. I would have loved to have evidence, even shaky, nonconclusive evidence, that the ghost of James L. Chaffin was real. Next I relate Sperry's theory about the brothers confronting Susie Chaffin over lunch. In repeating it, the story sounds hopelessly oversophisticated for a bunch of dirt farmers' sons. And why would they bother with the ghost, the overcoat, the slip of paper? Why not simply claim to have found the second will in the Bible?

"Well," says Lester, toeing an upended flower pot. "It's hard for me or you or anybody else to try to interpret the few facts we've got." I think he means, I wish you and your fancy-pants forensics man would stop trying. But he's too polite for that.

The car is quiet on the drive back to Mocksville. It's late on a Saturday afternoon. Families are sitting in kitchens and on porches, trading gossip, shucking corn, shooing flies. Tomorrow the churches will fill with men and women who hold no doubts about the existence of the human soul and its joyous postmortem journey, men and women who could not care less about the hog sausage opinions of a forensic document examiner and a writer from California. To them, these things are simple and certain: The Chaffins are honest folk. The soul is real. Flossie Gobble lives on.

Alas, for me, a belief is not something you are born into or that you simply choose to adopt one day. Belief, for me, calls for plausibility. And so I continue my wanderings. I have one more stop: a research venture taking place at the University of Virginia. I have saved this for last, because it represents what I think is my best chance for a speck of evidence that people leave their bodies when they die.

12

Six Feet Over

A computer stands by on an operating room ceiling, awaiting near-death experiencers

ON THE FAR WALL of an operating room in the University of Virginia Hospital is an enormous photograph of an alpine meadow. The sky and grass are the vivid, lit-up blues and greens of travel posters and ads for allergy drugs, and wildflowers are thick as snow. The beautiful scenery is intended to calm the surgical patients who come here. In the case of a patient I'll call Wes, the flowers have their work cut out for them. Wes is about to be momentarily—ever so briefly—almost killed.

The operation is a defibrillator insertion. Defibrillators are most recognizable as those electrified paddles you see being slapped on patients' chests during cardiac arrest scenes on *ER*. Nowadays they make defibrillators the size of cell phones and—if you're prone to dangerous heart arrhythmias—sew them right inside the chest.

The almost-killing is being done to test Wes's newly implanted defibrillator. An electrical charge will hit his heart at the crest of a specific EKG peak, derailing the beat and rendering the organ a quivering (fibrillating) lump of tissue incapable of pumping blood. With no oxygen being delivered to his brain, Wes will be clinically dead within seconds. (As long as a heart begins beating again within about four minutes, no permanent brain damage occurs.) It's then up to Wes's new defibrillator to jump-start the beat. Patients like Wes are ideal subjects for a study of near-death experiences.

Outside of the alpine panorama, Room 1 is a fairly standard operating room. There is the operating table, bulky and complicated. There is the towering bank of cardiac monitors, the anesthesiologist's station, the whiteboard on the wall ("21 Days to National Nurses Week!"). You would have to be looking carefully to notice anything out of the ordinary. It's up near the ceiling. Taped to the top of the highest monitor is an open laptop computer, as if perhaps they'd run out of study carrels over at the science library and were packing the students in wherever they could fit them. The computer belongs to Professor Bruce Greyson, who works a few blocks away, in the university's Department of Psychiatric Medicine.

Greyson has been studying near-death experiences (referred to by those who study them as NDEs) for twenty-nine years. It is difficult to sum up the NDE in a sentence. On a very nuts-and-bolts level, it's an experience in which a person who came close to dying recalls having been someplace other than blacked out inside his or her body. Some recall traveling no farther than the ceiling, rising away from themselves like a pocket of hot air; others remember hurtling through a sort of tunnel, often toward an all-encompassing light and sometimes toward family or friends* who have died. Patients who recall hovering near the ceiling sometimes report having watched their operation or resuscitation from above. Though their descriptions can be remarkably detailed and accurate (more on this later), some people argue that the patients might have been extrapolating from things they heard or felt, or unconsciously incorporating memories of TV medical dramas or previous hospital visits.

Greyson is trying to find out: Were they up there or not? In a study begun in early 2004, he hopes to interview eighty defibrillator insertion patients just after they come out of anesthesia. If they mention a near-death experience that included an out-of-body experience, he will ask them to describe everything they saw from up above. Appearing on Greyson's flat-open laptop during the operations is one of twelve images, in one of five colors, randomly selected by a computer program. The objects depicted are simple and familiar—a frog, a plane, a leaf, a doll. They are brightly colored and animated to help attract the patient's eye (or whatever it is you use to see when you've left your visual cortex behind). It's an ingenious setup: Since the laptop's screen faces the ceiling, the images can't be seen from below.

I rarely get excited about parapsychology experiments, but if this one produces even a single person who accurately describes the image, I'll be up there on the ceiling, too. So far, none of the subjects interviewed has reported any type of near-death experience. Working against Greyson is the cocktail of anesthesia used on the patients; it includes a drug that inter-

*Or occasionally, ex-husbands. A celebrity website reports that Elizabeth Taylor saw Mike Todd during her near-death experience. "He pushed me back to my life," she is quoted saying. Whether this was done for her benefit or his was not clear.

feres with their memory of anything they might experience (pain, fear, a field trip to heaven) while they're under. "Though if the consciousness is leaving the brain, then would memory matter?" mused Greyson as we walked here today. He shrugged. "I don't know." In a similar study four years back—done at Southampton General Hospital in England by cardiologist Sam Parnia and neuropsychiatrist Peter Fenwick—only four of sixty-three cardiac arrest survivors interviewed recalled a near-death experience and none reported seeing things from an out-of-body perspective.

Greyson is working in tandem with a team of UVA cardiologists led by Paul Mounsey. (Mounsey declined to speak with me.) Interestingly, cardiologists—not parapsychologists—have published some of the most widely read studies on near-death experiences. A notable example was the study by Dutch cardiologist Pim van Lommel, published in the *Lancet* in 2001. His primary aim was simple, if ambitious: to find out what causes the near-death experience.

Theories abound. Oxygen deprivation and the drugs used in anesthesia are commonly suggested, and indeed, both drugs and lack of oxygen can trigger elements of the near-death experience—including the tunnel and the light and the out-of-body experience—when death is not near. (Pot, hash, LSD, ketamine, mescaline, and fighter pilot training blackouts have all been known to induce NDE-like experiences.) Intense stress or emotional states have been cited, as have endorphins and seizures. And then there's the theory Greyson is testing for: the preposterous, marvelous, mind-whirling possibility that the patient's consciousness somehow exits, and operates independently of, his body.

Van Lommel and his team interviewed 344 cardiac arrest patients in ten Dutch hospitals. All the patients had been clinically dead (defined by fibrillation on their EKG), and all interviews were done within a few days of the resuscitation. Eighteen percent reported at least one aspect of the typical near-death experience. Van Lommel marvels at the medical paradox of the cardiac arrest NDE: Consciousness, perception, and memory appear to be functioning during a period when the patient has lost, to quote van Lommel, "all functions of the cortex and the brainstem. . . . Such a brain would be roughly analogous to a computer with its power source unplugged and its circuits detached. It couldn't hallucinate; it couldn't do anything at all."

The fact that only eighteen percent of resuscitated patients have any type of near-death experience led van Lommel to rule out medical explanations such as lack of oxygen to the brain. "With a purely physiological explanation such as cerebral anoxia . . ." he wrote, "most patients who have been clinically dead should report one."

Van Lommel found that his subjects' medication was statistically unrelated to their likelihood of having a near-death experience. (On the topic of anesthesia as an NDE inducer, Bruce Greyson makes the point that people under anesthesia but not close to death have far fewer NDEs than people who come close to death without being under anesthesia; so, as he puts it, "it's hard to see how the drugs can be causing the NDE.")

Fear was also unrelated to frequency of NDE (as was religious belief, gender, and education level). One of the explanations left standing was the last explanation you'd expect to read about in a copy of the *Lancet*: that perhaps the near-death experience was, to quote van Lommel's paper, a "state of consciousness . . . in which identity, cognition and emotion function independently from the body, but retain the possibility of nonsensory perception." Van Lommel ended his paper by encouraging researchers to explore, or at least be open to, the possibility that the explanation for NDEs is that the people having them are undergoing a transcendent experience. That is to say, their consciousness exists in, as van Lommel described it in a more recent paper, some "invisible and immaterial world."

Greyson and Mounsey are exploring it. It took some doing. The hospital's human subjects committee was uncomfortable with the study. To avoid upsetting his subjects, Greyson was asked to remove the word "death" from the consent forms and study title, a tricky undertaking when your study is on near-death experiences. Bear in mind, these are people with life-threatening heart conditions, people who are entering the hospital to *have their hearts stopped*. Greyson smiles. "And now for the dangerous part: I'm going to ask you if you remember anything."

We're back in Greyson's office, on the first floor of a creaky, converted Charlottesville house with a wide, inviting porch that no one has time to sit on. Greyson squeezes his near-death research in amid his teaching duties and his private psychiatric practice. I frequently get office e-mails back from him when it's 9 p.m. in his time zone. I'm not sure whether he has a family. On a shelf at his other office, at the hospital, there is a framed photograph of a child and another of some goats. "Is this your little girl?" I had asked him. He said no. I didn't know what to say next. "Are these your goats?" is what I came up with. He explained that he shared the office. Greyson is dressed today in a deep green button-down shirt and casual dress pants. He wears wire-frame glasses and an even brown mustache. His hair sits neatly on his head, and his hands rest mainly in his lap. There's a single barbell in the corner under a cabinet. I try, and fail, to picture him using it. Not that he seems unathletic. I just don't envision him in motion. I envision him sitting. Working. Working and working.

We've been talking about the stigma of parapsychology.

The University of Virginia is one of only three American universities with a parapsychology research unit or lab. Do they ever regret it? Greyson says there was a fair amount of debate as to whether to accept the original gift with which the parapsychology unit was founded. In 1968, Xerox machine inventor Chester Carlson, upon his wife's urgings, bequeathed a significant number of his millions to the University of Virginia for research on the question of survival of consciousness after death. The university seems to have made peace with their decision, and with the department. "Though if you talk to individuals," Greyson says, "you get the whole spectrum. Some people think this research is a waste of time and resources, and others think it's a valuable contribution to medical science." Though Greyson probably gets more respect from his parapsychology colleagues than from his peers in psychiatry, he seems to be held in high regard as a researcher here. On his mantel is a bronze bust—the university's William James Award for best research by a resident. I had never realized how much William James looks like Thomas Jefferson.

"That *is* Thomas Jefferson," says Greyson. "That's the only bust you can get in Charlottesville, Virginia."

THE FIRST CARDIOLOGIST to get involved in NDE research was Michael Sabom, currently in private practice in Atlanta. Sabom had read the work of psychologist Raymond Moody, Jr., who coined the term "near-death experience" and presented a series of cases in a 1975 book entitled *Life After Life*. Sabom was intrigued but skeptical. He was dissatisfied with Moody's anecdotal approach and the fact that no attempt had been made to independently verify the things that people had reported seeing while seeming to be outside their bodies.

Sabom, then a professor of medicine and cardiology at Emory University in Atlanta, decided to do a study of his own, a controlled study. Of 116 cardiac arrest survivors he interviewed, he found six who could recall specific medical details they'd seen during their near-death out-of-body experience. The six patients' descriptions of what they'd observed during their resuscitation were then compared to the report of the incident in their medical file. In no instances did the medical report contradict statements in the patient's description. Nor were there any medical errors.

This was not the case with Sabom's control group. Curious to see whether any old heart patient could come up with a convincingly detailed description of a cardiac resuscitation, Sabom interviewed twenty-five people who had spent time in coronary care units under similar circumstances to those of his subject group. All of them were familiar with the visuals of cardiac emergency: EKG monitors, defibrillator paddles, IV poles, crash carts. The controls were asked to describe, in as much detail as possible, what they would expect to see if their heart stopped beating and hospital staff attempted to resuscitate them. Twenty-two of the twenty-five descriptions contained obvious medical gaffes. Defibrillator paddles were hooked up to air tanks or outfitted with suction cups. The imaginary doctors were punching patients in the solar plexus and pounding on their backs instead of their chests. Hypodermic needles were being used to deliver electric shocks. It was as though chimps had been let loose in the emergency room.

Below is a passage from Sabom's interview with one of the six NDE patients who'd described the specifics of their resuscitations. It is fairly representative of the level of detail and seeming cohesiveness of these people's memories:

Where about did they put those paddles on your chest?
Well, they weren't paddles, Doctor. They were round disks with a handle on them. . . . They put one up here, I think it was larger than the other one, and they put one down here.

Did they do anything to your chest before they put those things on your chest?
They put a needle in me. . . They took it two-handed—I thought that was very unusual—and shoved it into my chest like that. He took the heel of his hand and his thumb and shot it home. . . .

Did they do anything else to your chest before they shocked you?
Not them. But the other doctor, when they first threw me up on the table, struck me. . . . He came back with his fist from way behind his head and he hit me right in the center of my chest. . . . They shoved a plastic tube like you put in an oil can, they shoved that in my mouth.

Another patient describes a pair of needles on the defibrillator unit, "one fixed and one which moved," which was typical of 1970s-era defibrillators. (The man's heart attack happened in 1973.) Sabom asks him how it moved, to which he replies, "It seemed to come up rather slowly, really. It didn't just pop up like an ammeter or a voltmeter, or something registering. . . . The first time it went between one-third and one-half scale. And then they did it again, and this time it went up over one-half scale." Though the man had been an air force pilot, he had never seen CPR instruments during his training.

Of course, it's possible Sabom's subjects were extrapolating from things they'd felt and heard, either just before their heart stopped or some time afterward. (The interviews were done years after the incidents had taken place, so doctors couldn't be relied upon to verify the timing of specifics.) It's possible the patients could have heard what the doctors and nurses were saying and subconsciously fabricated visual details to match. Hearing is the last sense to disappear when people lose consciousness. Dozens of articles have run in medical journals over the years addressing concerns about anesthetized patients hearing the things said about them during surgery.* Not just things like, "Nurse, more suction." Things like, "This woman is lost" and "How can a man be so fat?"—both actual examples reported by patients in a 1998 *British Journal of Anaesthesia* article.

If it's possible the patients heard things, it's also possible they might have partway opened their eyes and seen things. And the things they saw could then have been incorporated into the viewpoint of being up above the scene. A couple of years back, epilepsy researchers at the Program of Functional Neurology and Neurosurgery at the University Hospitals of Geneva and Lausanne stumbled onto a site within the brain that, when stimulated, reliably caused the perception of looking down on one's body from above. So convincing were the images that the patient in question pulled back when asked to raise her knees, because it appeared to her that her knees were

---

*My favorite being "The Anesthetized Patient *Can Hear* and *Can Remember*," from a 1962 *Journal of Proctology* article. "Their physiologic adaptations to the stress of surgery may be profoundly disturbed by what they hear," wrote the author, leading me to mistake him for a caring physician. Then he went on: "Medico-legal implications are obvious even if we do not care about the patient." I sat there blinking in disbelief. I did this again twelve pages later, upon seeing the emblem of the International Academy of Proctology: a double-snake caduceus with a free-floating length of rectum standing in for the pole.

about to hit her in the face. The visuals were limited to the person's own body, however, and not the furniture or equipment or researchers around it. Still, one can imagine a blending of this viewpoint with information gleaned from things heard or seen.

The holy grail of NDE research, then, the best evidence that what seemed to be an extrasensory perception was indeed extrasensory, would be a deaf and blind patient: someone who "sees" things during a near-death experience that are later verified and that couldn't have been inferred from something he or she saw or heard—because he or she can't see or hear.

The closest Sabom has come to this is a woman named Pam Reynolds, who, in 1991, underwent brain surgery with her eyes taped shut, and molded, clicking inserts inside her ears. (Watching the brain stem's responses to clicks is a way of monitoring its function.) Despite this, and despite the fact that her EEG was flat, meaning all brain activity had stopped (surgeons were repairing a massive aneurism and had drained the blood from her brain), she reported having "seen" the Midas Rex bone saw being used on her skull. She said it looked like an electric toothbrush and that its interchangeable attachments were kept in what looked like a socket wrench case. I went on the Midas Rex web page to have a look at their bone saws. Indeed, bone saws look nothing like any saw I've ever seen. They do look like electric toothbrushes—not the kind you or I might use, but the kind dentists use, with interchangeable heads and a metal handle attached to a long flexible tube that leads to a motor housing. After I'd recovered from reading the copy ("true high-speed bone-dissecting performance"! . . . "For cutting, drilling, reaming . . ."), I clicked on the Instrument Case page, where the various attachments were shown in a box resembling nothing so much as a socket wrench case.

But why was Reynolds unable to describe any of the people in the room? Sabom nominates "weapon focus phenomenon," which you can read all about in a 1990 issue of the *Journal of Law and Human Behavior*. Research has shown that victims of armed criminals are able to accurately recall the weapon used on them ninety-one percent of the time, and the guy holding it only thirty-five percent of the time. So perhaps the bone saw had hijacked Pam Reynolds's attention. Or, who knows, perhaps she paid a visit to the Midas Rex web page, too. This is the trouble with anecdotes.

Though there is no deaf-blind NDE study, there is a study of blind people who have had NDEs. Psychology professor and International Association for Near-Death Studies cofounder Kenneth Ring and then–psychology Ph.D. candidate Sharon Cooper contacted eleven organizations for the blind, explaining that they were looking for blind people who had had near-death or out-of-body experiences. They ended up with thirty-one subjects (and a book, called *Mindsight*, published in 1999). Twenty-four of these subjects reported being able to "see" during their experiences. Some "saw" their bodies lying below them; some "saw" doctors or physical features of the room or building they were in; others "saw" deceased relatives or religious figures.

Strangely, the subjects who reported "seeing" these things included people who had been blind from birth: individuals whose dreams almost never contain visual images, just sounds and tactile impressions. An example is a man named Brad, who reported having floated up above the building, where he could see snowbanks along the streets, of "a very soft kind of wet" slushy snow. He saw a playground and a trolley going down the street. When asked if perhaps he did not see but somehow sensed these things, Brad replied, "I clearly visualized them. I remember being able to see quite clearly." (Others were less decisive: "It was seeing but it wasn't vision," said a woman named Claudia.) Understandably, the experience was confusing and, in one woman's words, frightening. "It was like hearing words and not being able to understand them," she told Ring, "but knowing they were words."

I was mainly interested in whether any specific, unique details of what the blind people had "seen" could be verified by others who had seen these details, too. The book includes a chapter on corroborative evidence, but it is a bit disappointing. Often the people who could have verified what the blind people said they'd seen were impossible to track down, or did not recall any details of the events. One exception was a woman named Nancy, who lost her sight as a result of surgical complications. (They accidently cut and then *sewed shut* a large vein near her heart.) After the mishap, on her way into emergency surgery, she "saw" both her lover and the father of her child standing down the hallway from where her gurney was being wheeled toward an elevator. Ring tracked down both the lover and the dad, and both confirmed that they had watched her gurney go by from down the hall. However, there was some question as to exactly when she had gone blind (i.e., was it before or after the gurney ride?). And it's hardly the kind of whiz-bang dazzle shot—to borrow Gary Schwartz's terminology—that you hope for. You'd want the two men, or at least one of them, to have been "seen" (and then verified by someone else) doing something unique, something other than just being there—eating a banana, say, or tripping over an IV pole.

The most impressive near-death dazzle shot I've come across was not something reported by a blind person. It was a sneaker, seen by a migrant worker named Maria, who had a heart attack in Seattle. Maria told her ICU social worker—a woman whose parents did her the gross disservice of naming her Kimberly when her last name was Clark*—that she had not only spent time watching herself being worked on by the ER team, but had drifted out of the building and over the parking lot. It was from this perspective that she noticed a tennis shoe on a ledge on the north end of the third floor of the building. Later that day, Kimberly Clark went up to the third floor and found a tennis shoe where Maria had reported seeing one. Unfortunately, she didn't bring along a witness.

The sneaker story eventually made its way to Kenneth Ring. In much the same way as unverified anecdotes of blind people's near-death "sights" prompted his *Mindsight* study, Ring set out in search of other "cases of the Maria's shoe variety," cases he would then attempt to verify. He found three, which he describes in a 1993 article in the *Journal of Near-Death Studies.* Oddly, two of the three incidents involve shoes. In the first anecdote, Ring communicated with an ICU nurse who had returned to work from vacation wearing a new pair of plaid shoelaces. A woman she helped resuscitate saw her the next day (presumably in a different pair of shoes) and said, "Oh, you're the one with the plaid shoelaces." When the nurse expressed surprise, the woman said, "I saw them. I was watching what was happening yesterday when I died." Another out-of-body heart attack patient reports to a nurse that he saw a red

*In checking the spelling of "Kimberly-Clark" on the web, I note that the personal hygiene empire has expanded well beyond sanitary napkins. It's a global powerhouse spewing forth multiple brands of diapers, adult diapers, disposable training pants, bed-wetting underpants, "flushable moist wipe products," award-winning disposable swim pants, and "cloth-like towels strong enough for big messes," though probably not the big mess of umpteen billion used disposable hygiene products.

shoe on the hospital roof; a skeptical resident gets a janitor to let him up onto the roof, where he finds a red shoe (and loses his skepticism). No doubt someone out there is working on a journal article about "shoe focus phenomenon," but until then, the out-of-body traveler's affinity for footwear must remain a mystery.

Ring interviewed both these nurses, though apparently could not track down any third parties to corroborate the stories. It's possible the patients had somehow seen these items before surgery. It's also possible, in the case of the shoe on the roof, that it's a coincidence. You can't be sure. You're relying on one person's claim. The danger of that is best expressed in the form of a hand-glued last-minute errata slip in Ring's book:

> Readers are advised to disregard entirely the . . . Appendix, in which a case of a blind woman who purported to have an NDE is described. . . . We discovered, to our chagrin, that this case has fraudulent aspects. Dr. McGill, who offered this account to us in good faith, now believes she was deceived by the woman in question.

That's why I like the computer-near-the-ceiling project. It's a study, not an anecdote. Unfortunately, it's a slow-moving study. Because of limitations imposed by the human subjects committee, Greyson has interviewed fewer than thirty subjects to date.

Is there any other experimental avenue for proving that a mind (soul, personality, consciousness, whatever) can travel independent of its body? There is, though it's not an avenue along which mainstream researchers would be willing to stroll. It involves people who claim to be able to will themselves to have out-of-body experiences—simply pull their consciousness out of the garage and take it for a ride.*

If you wanted to prove that it's possible for some version or vestige of the self to exist independent of body and brain, you could try to set up some sort of detector in a room far away from one of these purported free-floaters, and instruct him or her to head on over. It's a jump to further conclude that this is what we do when we die, but it would make it easier—for me, anyway—to accept that NDEs are something other than a neurological/psychological phenomenon.

In 1977, a group of parapsychologists undertook just such a project, on the campus of Duke University. I was pleased to see that the main author on this study was the late Robert Morris, of the University of Edinburgh. I'd written an article on Morris's telepathy work years ago; I liked the fact that he had cooperated with the skeptics group CSICOP (Committee for the Scientific Investigation of Claims of the Paranormal) in designing the experimental protocol.

Morris and his colleagues worked with a single subject named Stuart Harary, who had participated in previous out-of-body experience projects at Duke. Harary was instructed to leave his body and travel to one of two detection rooms, either fifty feet or a quarter mile away. To determine whether he could actually do this, Morris stationed people in the detection room and had them try to sense Harary while he "visited." The results were no better than chance. There were about as

---

*There is, of course, disagreement as to whether they are actually traveling somewhere or simply experiencing a vivid hallucination; a good discussion of this can be found in the *Skeptical Inquirer* article by Susan Blackmore listed in the bibliography. Blackmore, a parapsychologist turned skeptic, has had out-of-body experiences of her own, which you can read about on the website of TASTE, The Archives of Scientists' Transcendent Experiences.

many reports of detection during control periods as when Harary believed he was out of his body.

Surmising that animals might be more keenly attuned to extrasensory presences, Morris next did a series of trials using snakes, gerbils, and kittens as detectors. The cages were set up on top of an activity platform that registered movements on a polygraph, whose readout could then be compared to the timing of Harary's "visits." As anyone who's been to a herpetarium could have predicted, the snakes did not work out. They didn't move around when Harary was visiting the room, and they didn't move around when he wasn't. The gerbils proved similarly apathetic. "The rodents spent most of their time either chewing on the cage bars or resting quietly," wrote Morris.

Morris eventually settled on a kitten that had seemed to show an affinity for Harary. The kitten was not caged but let loose in a corralled area with a grid taped out on the floor; in this case the behavioral measure was the number of squares entered per one hundred seconds and the animal's vocalization rate. Disappointingly, the kitten seemed to be reliably less active when Harary indicated he was "there," leading some of the researchers to wonder whether they'd gotten the protocol backward. Perhaps Harary's presence wasn't stimulating the animal but calming it. Morris and his colleagues went through a half dozen methodological variations, including one in which the kitten was sequestered under an inverted box until Harary "arrived," whereupon the box, which was hooked to a pulley, would slowly and dramatically rise like a stage curtain. It is around this point that I like to insert the image of a group of white-haired Duke alumni wandering into the building on a homecoming tour.

The experiment dragged on so long that around page 11, Morris begins referring to the kitten as a cat, noting that it had by then grown to maturity. He reported a number of anecdotal occurrences—frustratingly, a couple of casual bystanders proved better at sensing Harary than the official "human detectors"—that would seem to indicate something was up, but overall there was little to suggest that Harary had been anywhere but inside Harary's head. Nonetheless, everyone seemed to have a good time, and scientific literature is the richer for the introduction of the measurement unit "meows per second."

A few years later, a team of non-university-affiliated paranormal researchers tried a similar experiment, with strain gauges in place of kittens and gerbils. Here our out-of-body traveler was an amateur parapsychologist from Maine named Alex Tanous. For clarity, Mr. Tanous referred to his out-of-body self as Alex 2, and his stay-at-home self as Alex 1, and so I will, too. Alex 2's mission was to travel six rooms distant, enter a suspended (to keep floor vibrations from setting off the strain gauges) eighteen-inch cube, and view one of five randomly generated images, which would appear in one of four colors and four quadrants. Meanwhile, Alex 1 would tell the researchers what he sees. A tape was kept running, so that the researchers could see if the strain gauges were registering force specifically when Alex 2 was correctly reporting what he "saw," as this would suggest that he had actually been inside the cube—rather than knowing the target remotely, via some more ordinary, garden-variety ESP.

Head researcher Karlis Osis, who died in 1997, reported that Tanous had 144 hits and 83 misses. Does this mean that Tanous got all three aspects (color, quadrant, image) of the target correct sixty-three percent of the time? When chance would dictate a correct guess only once in eighty tries? Why hasn't this guy been on the news? Why hasn't he turned the world upside down? Osis further claimed that when Alex 2 was seeing the targets correctly, the mean activation level of the gauges inside the chamber was significantly greater than when he wasn't seeing the target correctly. "Therefore," concludes the paper, "it is our opinion that the [strain gauge] results can most likely be attributed to the subject's out-of-body presence in the shielded chamber."

Though I suspected that a conversation with Tanous would leave me chewing on my cage bars, I decided to try to call him. I did not succeed, because in 1990, Alex 2 had, like Osis, made the big one-way trip out of his body. So we are left to conclude that either Tanous was some sort of bizarre on-call living ghost, or Osis was a deluded or sloppy researcher.

SO LET'S SAY, just for a moment, that people who have near-death experiences are actually leaving their bodies. That they are making some sort of transcendent journey into a different dimension. And that one of the off-ramps in this dimension leads to the afterlife. This means that near-death experiences could—just possibly—provide us with a sneak preview of our own impending eternity. If only someone had kept a list of near-death experiencers' descriptions of this place.

Someone did. Michael Sabom's book includes an appendix of all twenty-eight "transcendental environments" glimpsed or "visited" during subjects' near-death experiences. There seem to be two basic versions: the weather report and the farm report. Fully half the environments consisted of nothing but sky. Heaven appears to have a similar weather system as earth; there were approximately the same number of reports of blue sky and sunshine as there were of clouds and mist. One or two meteorologically inclined individuals included both in their report (e.g., "blue sky with an occasional cloud").

The other half of the twenty-eight descriptions consisted of gardens or pastures, often with a gate thrown in. The heavenly farmland was more or less deserted, the exceptions being one pasture with cattle grazing and one landscape of people "of all different nationalities, all working on their arts and crafts."

It seems pretty clear what's going on here. People are experiencing something dazzling and euphoric and totally foreign, and interpreting it according to their image of heaven. Greyson agrees. "I think the experience is so ineffable that we just put whatever framework, whatever models and analogies we have, onto it." Greyson says these cultural overlays also apply to the experience of rushing down a tunnel. "I had one truck driver I interviewed call it a tailpipe." Likewise the experience of being sent back to return to one's body. In one journal article I read, a man who lived in India experienced this as being told there'd "been a clerical error."

The alternate explanation, of course, is that the people who had these NDEs actually saw heaven, and that heaven looks just like it looks in the holy books. This is, of course, tricky to prove. Someone who's been there would have to bring back photographs. Preferably someone scientific, someone trustworthy and pedigreed.

> On December 26, 1993, the Hubble Telescope made visual contact with Heaven and took hundreds of pictures and sent these pictures of Heaven to Goddard Space Center in Maryland. . . . In the pictures of Heaven, you can see bright light and what looks like the Holy City. . . . Heaven is located at the end of the Universe.

This dispatch comes to us courtesy of the Internet Religious News service. One fine day I called Goddard Space Flight Center to see what they had to say about this. "Well," said a good-natured NASA spokesperson named Ed Campion, "it is true that Hubble focuses on faint lights at the most distant parts of the universe." That's why NASA sent a telescope way out into space—to get it closer to the oldest, most distant parts of the universe, the stuff that dates to the Big Bang. But Campion hadn't heard about the heaven photos. Or the secret NASA space probe that recorded millions of voices singing "Glory, Glory, Glory to the Lord on high" over and over—as reported, here again, by our imaginative friends at the Internet Religious News service. Or the NASA photos of the "two Giant Human-Looking Eyes in deep space that are billions of light years around and billions of light years apart looking at Earth."

"That last one," said Campion, "kind of gives me the willies."

Realistically speaking, if the place experienced by people who almost die exists as something other than a neurological phenomenon, it's no more likely to be located by astronomers than the soul was likely to be located by the early anatomists. It exists (if it exists) in a dimension other than that of time and space, a dimension we typically can't access (if indeed we do access it) until we die. Greyson's thinking is that cessation of the brain's everyday activities, as happens during clinical death, might enable the consciousness to tune in to a channel normally blocked or obscured by the chatter. "It's almost as if the brain in its normal functioning stops you from going there," he told me. "And when you knock those parts of the brain out, then you're able to."

Other than a brush with death, are there other ways to deflect that bothersome everyday sensory input and experience the transcendent reality that may or may not be out there all the time? You bet. The following is a passage from a chapter on the drug ketamine in the book *Anesthesiology*: "I would suddenly find myself going down tunnels at high speed. . . . One time I came out into a golden light. I rose into the light and found myself having an unspoken interchange with the light, which I believed to be God." London psychiatrist and ketamine authority Karl Jansen quotes the passage in his own book *Ketamine: Dreams and Realities.* Ketamine is today rarely used as an anesthetic and fairly commonly used as a recreational drug. Jansen used to be of the opinion that since ketamine—or LSD or pot, for that matter—can produce ersatz near-death experiences, this meant that surgical or cardiac arrest patients' near-death experiences were similarly hallucinogenic.

He has of late changed his tune: "The fact that near-death experiences can be artificially induced does not imply that the spontaneously occurring NDE is 'unreal' in some way," writes Jansen in *Ketamine.* "It has been suggested that both may involve a 'retuning' of the brain to allow the experience of a different reality from the everyday world." If this is true, it suggests it may be possible to preview death by taking ketamine—which is precisely what self-described psychonauts Timothy Leary and John Lilly did, in what they called "experiments in voluntary death."

If you want to have a K-induced near-death experience, I read in Jansen's book, you should take a fairly ambitious dose, and you should take it by injection. You should also be prepared for all manner of physical side effects ranging from the dangerous to the embarrassing. Your eyes may wander off in different directions. Your body may jerk uncontrollably. In his book, Jansen passes along the advice of *The Essential Psychedelic*

Guide author D. M. Turner, which is to have a friend or "sitter" present when you take ketamine. (Turner, Jansen wryly points out, died alone in the bath with a bottle of K beside the tub.)

Before I traveled to England for my mediumship course, I scheduled an interview with Jansen in London. I wanted him to be my sitter, but I didn't, at this point, tell him. I was hoping he'd offer. I imagined we'd sit in his office and chat for a while, and then he'd open a drawer. *I happen to have some ketamine right here. I'll gladly provide you with a safe, clinical environment in which to have a near-death experience with absolutely no unsightly side effects.*

Jansen made no such offer. He was in the process of relocating to New Zealand and was staying at a hotel while in London. He had no office, so we met in his hotel bar. He was tall and suave and accompanied by a similarly tall and suave Russian woman. We talked for half an hour, shouting over the din while the Russian woman looked on intently. I imagined her looking on intently while my eyes rolled around in my head and my extremities spazzed. I no longer wanted to take drugs with Dr. Jansen.

The clean-and-sober voluntary-death alternative takes the form of a Buddhist meditation. Pure Land Buddhists, who date back to A.D. 400, believe that by practicing certain rather extreme forms of meditation, it's possible to experience the same heavenlike locale that people report having experienced during brushes with death. These guys were the original near-death experience researchers. One of the junior monks' duties was to sit at the bedside of moribund elders and jot down their deathbed visions of the Pure Land. By the eleventh century, more than a hundred accounts of the Pure Land had been transcribed, including many from people who had been thought dead and then revived. The monk Shan-tao was one of the most ardent devotees; his sermons contained long, vivid descriptions of the Pure Land. Possibly *too* vivid: Osaka University professor Carl Becker writes in an article in *Anabiosis: The Journal for Near-Death Studies* that at least one listener was compelled to take the express route to the Pure Land, committing suicide in the days following the sermon.

Should you, too, wish to preview the afterlife, here are instructions for the "constantly walking meditation practice": "For a single period of 90 days, only circumambulate exclusively. . . . Until three months have elapsed, do not lie down even for the snap of a finger. Until the three months have elapsed, constantly walk without stopping (except for natural functions)."

Before you begin, I should warn you that both Pure Land Buddhists and ketamine users occasionally experience something closer to hell than heaven. As do near-death experiencers. Researcher P. M. H. Atwater, who interviewed more than 700 people about their near-death experiences, reported that 105 of these individuals described their experience as unpleasant. But only one researcher ever claimed to be hearing tales of literally hellish goings-on. Cardiologist Maurice Rawlings recounted dozens of stories of people hearing screams and moans and witnessing violent scenes of gruesome torture at the hands of grotesque animal-human forms. Rawlings raised eyebrows in the NDE community with his second book, which advocated a commitment to Christianity as a way of ensuring one doesn't end up in the sorts of hellish scenarios he claimed his non-Christian near-death experiencers were describing.

If you take Rawlings out of the picture, reports of hell-like sights and sounds are rare.* You will be pleased to know that Atwater never once heard a description of a fiery or even unseasonably hot locale. Both Atwater and Greyson concluded that the difference between an unpleasant near-death experience and a pleasant one is largely one of attitude. A bright light at the end of a tunnel can seem warm and inviting, or it can seem mysterious and terrifying. People of the world "all working on their arts and crafts" can seem like heaven or, if you're me, hell. The same vast expanse of empty sky that looks beautiful to one person may seem lonely and barren to another. I once interviewed a geologist who searches for meteorites on empty, wind-battered ice fields in Antarctica, where the snow is whipped into knee-high white swirls. He sometimes gives talks and slide shows of his travels to the public. Most people tell him they can't imagine spending months at a time in such a cold, barren locale. One night a quiet older woman came up to him as he was putting away his slides and said, "You've been to heaven."

Bruce Greyson has also written papers on what he calls the distressing near-death experience. I asked him whether researchers had ever looked for a correlation between having a hellish near-death experience and being a mean, rotten person. Just, you know, wondering. His answer was reassuring: "We have very blissful accounts from horrible people." He told me

---

*That is, in the near-death journals. You can find them in certain fundamentalist Christian publications. I read that in the February 1990 issue of the Trinity Broadcasting Network newsletter *Praise the Lord*, there's an article about scientists drilling in Siberia and suddenly poking through to a hollow space from which issued screams and temperatures in excess of two thousand degrees. I spoke to a woman in the newsletter department at TBN, who apologized for not being able to send out pre-2003 back issues. "We disregard them every year," she explained confusingly. "We shred them."

the story of a Mafia bagman who was shot in the chest and left to die. While lying there bleeding, he had "a beautiful experience, in which he felt the presence of God and unconditional love." One of the focuses of Greyson's near-death work has been the effects—often profoundly positive—that near-death experiences have on people's lives. The bagman, for example, quit the Mafia and now counsels delinquent boys. "He walked away from his lifestyle," says Greyson. "I talked to his former girlfriend, who used to complain to me: 'Rocky just doesn't care about money, about things of *substance* anymore.' "

I DON'T KNOW if Wes can hear anything, but he surely can't see. His face, like the rest of his body, is draped in blue surgical cloths. If he could see, he'd surely be entertained. Everyone in the room is dressed in bulky lead kilts and matching lead dickeys to protect their thyroids and reproductive organs from the real-time X-rays that are helping the surgeons thread a sensor wire through Wes's* heart. The wire will be connected to the body of the defibrillator, soon to be sewn into a pocket in the pectoral muscle just below the spot where a more conventional shirt pocket would be.

And now it's time to almost kill Wes. A technician from the defibrillator company fiddles with a small computer that

*And now I must reveal to you that Wes is not a defibrillator insertion patient in Charlottesville, but in San Francisco, near where I live. The human subjects committee for Greyson's study would not allow me in the operating room. So I called UCSF Medical Center, who kindly let me observe an insertion. My apologies to the reader, and my thanks to UCSF Medical Center (number six on *U.S. News & World Report*'s 2004 list of the nation's best hospitals). And to the unconscious Wes, who later wrote and apologized for "not having been more sociable."

remotely manipulates the implanted device. In the corner of the screen is the company's disquieting logo, a heart with a jagged lightning bolt through it. "We're preparing to shock," announces the technician. Depending on the voltage and on what the heart is doing when you shock it, the charge can either induce or stop fibrillation. "So it can kill him, or save him," she says brightly.

This time, they're aiming—temporarily, of course—for the former. "Here we go," says the technician. "I'm enabling and . . . I'm inducing." The jolt makes Wes's chest muscles contract violently, jerking his torso up off the table as though he'd been kicked from below. "We have VF," says the technician, sounding all urgent and mission-control. "VF" stands for ventricular fibrillation. On the EKG monitor, Wes's heartbeat dithers wanly. What's going on in his mind right now? Is he beholding the bright light? Speeding through the tunnel? Attending an appliqué class? Wherever he is, it's a brief visit; three seconds later the defibrillator is preparing to shock his heart back to lub-dub.

Twenty minutes later, Wes is being wheeled to the recovery room. Technically speaking, anyone who makes it to a recovery room can't have been dead. By definition, death is a destination with no return ticket. Clinically dead is not *dead* dead. So how do we know the near-death experience isn't a hallmark of dying, not death? What if several minutes down the line, the bright light dims and the euphoria fades and you're just, well, dead? We don't know, says Greyson. "It's possible it's like going to the Paris airport and thinking you've seen France."

Greyson is an inestimably patient person in a field rife with inconclusive data and metaphysical ambiguities. I ask him what he thinks, in his heart of hearts. Does the personality survive death? Surely, after all these years, he has an opinion. "It wouldn't surprise me at all if we come up with evidence that we do survive. I also wouldn't be terribly surprised if we come up with evidence that we don't."

Sabom is less equivocal. I asked him, in an e-mail, whether he believed that the consciousness leaves the body during an NDE and is able to perceive things in an extrasensory manner. "Yes," came the reply.

I asked van Lommel the same question, and got the same reply. "I am quite sure that it is not a hallucination or a confabulation," he wrote. "I am convinced that consciousness can be experienced independently from the body, during the period of a nonfunctioning brain, with the possibility of nonsensory perception."

Van Lommel mailed me a draft of a new article in which he presents a theory as to how this might be possible. He uses the analogy of radio or TV transmissions. All these channels, these different electromagnetic fields packed with information, are out there all the time. We can't watch HBO if we're already watching Bravo, but that doesn't mean HBO's broadcast ceases to exist. "Could our brain be compared to the TV set, which receives electromagnetic waves and transforms them into image and sound? When the function of the brain is lost, as in clinical death or brain death, memories and consciousness still exist, but the receptivity is lost, the connection is interrupted." Then he went all Gerry Nahum on me. His paper stepped into quantum mechanics, to phase-space versus real-space, to nonlocality and fields of probability. Neuronal microtubules made an appearance. I had to set it down.

I can't evaluate this sort of theorizing, because I have no background in quantum physics. A few months ago, I was corresponding with a Drexel University physicist named Len Finegold. I mentioned quantum-mechanics-based theories of consciousness. You can't hear someone sigh through e-mail, but I heard it anyhow. “Please beware,” came his reply. “There are a lot of people who believe that just because we don’t have an explanation for something, it’s quantum mechanics.”

So I’m holding out for the guys on the ceiling. As soon as someone sees an image on Bruce Greyson’s computer, you can mark me down as a believer.

Last Words

SOMEWHERE THIS past year, I read that the most powerful influences upon your opinion about paranormal phenomena are your friends and family. The closer you are to the teller of a ghost story, the more likely you are to believe that the ghost in the story was a ghost, and not a raccoon or a temporal lobe seizure. Your beliefs are formed not by researchers or debunkers or television psychics, unless perhaps one of them is your mother or your good pal. Your beliefs are formed by your own experiences and those of your inner circle. And then validated by the researchers or the debunkers or the television psychics.

Now that you've spent 291 pages with me, I suppose I almost fall into the category of a friend, or anyway, someone that you know. And you might be wondering what it is, at this point, I believe. Has my year among the evidence-gatherers left me believing in anything I didn't believe in a year ago? It has. It has left me believing something Bruce Greyson believes. I had asked him whether he believes that near-death experiences provide evidence of a life after death. He answered that what he believed was simply that they were evidence of something we can't explain with our current knowledge. I guess I believe that not everything we humans encounter in our lives can be neatly and convincingly tucked away inside the orderly cabinetry of science. Certainly most things can—including the vast majority of what people ascribe to fate, ghosts, ESP, Jupiter rising—but not all. I believe in the possibility of something more—rather than in any existing something more (reincarnation, say, or dead folks who communicate through mediums). It's not much, but it's more than I believed a year ago.

Perhaps I'm confusing knowledge and belief. When I say I believe something, I mean I *know* it. But maybe belief is more subtle. A leaning, not a knowing. Is it possible to believe without knowing? While there are plenty of people who'll tell you they know God exists, in the same way that they know that the earth is round and the sky is blue, there are also plenty of people, possibly even the majority of people who believe in God, who do not make such a claim. They believe without knowing. I remember once standing in the kitchen of my friend Tim, having a conversation about organic milk. I explained, in my usual overagitated, long-winded way, why I wasn't yet convinced of the need to part with an extra dollar a quart. I didn't *believe* in organic milk. Tim, who buys organic milk, listened to me for a while, and then he shrugged. "It's just a decision," he said. In other words, you don't have to go out and read every published paper on antibiotics and bovine growth hormone, weighing those that speak for milk's safety against those that warn of its dangers, before you can decide to believe in buying organic. You don't need proof. You just need an inclination.

Perhaps I should believe in a hereafter, in a consciousness that zips through the air like a *Simpsons* rerun, simply because it's more appealing—more fun and more hopeful—than not believing. The debunkers are probably right, but they're no fun to visit a graveyard with. What the hell. I believe in ghosts.

Acknowledgments

PEOPLE ASSUME that authors are experts in the field about which they have chosen to write. Possibly most are. Possibly I'm the only one who begins a project from a state of near absolute ignorance. But I do, and it makes me an especially irksome presence in my sources' lives. I ask naive, misguided questions and giggle at the wrong moments. I stay too long and grasp too little. The following names are listed in order of diminishing exasperation: Kirti Rawat, Bruce Greyson, Gerry Nahum, Gary Schwartz, Michael Persinger, Julie Beischel, Vic Tandy, Allison DuBois, Grant Sperry, and Karl Jansen, please accept my thanks for your patience and generosity and my apologies for the limits of my experience and the blind spots of my mindset.

For miscellaneous offerings of wisdom and arcane fact, a

formal bow to Jürgen Altmann, Peter Copeland, Marco Falcioni, Jürgen Graaff, Lew Hollander, Jr., Nan Knight, Greg Laing, Anne LeVeque, His Excellency Pasquale Macchi, Peggy Pearl, Dean Radin, Eric Ravussin, Colleen Phelan, Julie Rousseau, Michael Sabom, Pim van Lommel, and Valerie Wheat. A tip of the hat to Kim Wong, Susan Grizzle, and Wes Lange, who got me into the operating room and out of a logistical pickle; to everyone at the Grotto; and to the ever-miraculous interlibrary loan staff of the San Francisco Public Library.

Lester, Ruby Jean, and Lloyd Blackwelder must have their own paragraph, because they not only helped me and trusted me with their story, they practically adopted me. If I could bake, I'd send you a persimmon pie.

I hesitate to thank Jay Mandel as my agent, because that is only one of the many hats I force him to wear on my behalf: reader, advisor, hand-holder, career counselor. You make it all easy. Similarly indispensable guidance and good humor came from Jill Bialosky, who has the gall to be as gifted an editor as she is a writer. The two of you have taken me on an incredible trip, for which I am deeply, unabashedly grateful.

A book is a collective undertaking, and this one, like the last, benefited tremendously from the talents of Bill Rusin and the rest of the Norton sales staff, Deirdre O'Dwyer, Erin Sinesky, and Jamie Keenan, whose covers make my heart fizz.

And then there is Ed, to whom every mushy cliché applies and none does justice.

Bibliography

Chapter 1: You Again

Angel, Leonard. "Empirical Evidence for Reincarnation? Examining Stevenson's 'Most Impressive' Case." *Skeptical Inquirer* 18: 481–87 (Fall 1994).

Bertholet, D. Alfred. *The Transmigration of Souls.* Translated by Rev. H. J. Chaytor. London and New York: Harper & Brothers, 1909.

Hopkins, Edward W., ed. *The Ordinances of Manu.* London: Trübner, 1884.

O'Connell, Rev. J. B. *The Celebration of Mass: A Study of the Rubrics of the Roman Missal.* Milwaukee: Bruce Publishing, 1944.

Stevenson, Ian. *Twenty Cases Suggestive of Reincarnation*, 2d ed. Charlottesville: University Press of Virginia, 1980.

——. *Reincarnation and Biology*. Westport, CT: Praeger, 1997.

Tucker, Jim B. "A Scale to Measure the Strength of Children's Claims of Previous Lives: Methodology and Initial Findings." *Journal of Scientific Exploration* 14 (4): 571–81.

Chapter 2: The Little Man Inside the Sperm, or Possibly the Big Toe

Ackerknecht, Erwin, and Henri V. Vallois. *Franz Joseph Gall, Inventor of Phrenology, and His Collection*. Wisconsin Studies in Medical History, No. 1. Translated by Claire St. Léon. Madison, WI: Department of History of Medicine, University of Wisconsin Medical School, 1956.

Bailey, Percival. "The Seat of the Soul." *Perspectives in Biology and Medicine*, Summer 1959: 417–41.

Dobell, Clifford. *Antony van Leeuwenhoek and His "Little Animals."* New York: Harcourt, Brace, 1932.

Ford, Norman. *When Did I Begin?* Cambridge: Cambridge University Press, 1988.

Gall, Franz J. *Sur les fonctions du cerveau et sur celles de chacune de ses parties* . . . Paris: J. B. Baillière, 1825.

Grüsser, O. J. "On the 'Seat of the Soul': Cerebral Localization Theories in Mediaeval Times and Later." In *Brain—Perception—Cognition: Proceedings of the 18th Göttingen Neurobiology Conference*, edited by Norbert Elsner and Gerhard Roth. New York and Stuttgart: Georg Thieme Verlag, 1990.

Kaitaro, Timo. "La Peyronie and the Experimental Search for the Soul: Neuropsychological Methodology in the Eighteenth Century." *Cortex* 32: 557–64 (1996).

La Peyronie, F. G. "Observations par lesquelles on tâche de découvrir la partie de cerveau où l'âme exerce ses fonctions." In *Mémoires de l'académie royale des sciences*, 1741, pp. 199–218. Paris: Chez G. Martin.

Leeuwenhoek, Antoni van. *The Select Works of Antony van Leeuwenhoek.* Translated by Samuel Hoole. London: G. Sidney, 1800.

Peacock, Andrew. "The Relationship Between the Soul and the Brain." In *Historical Aspects of the Neurosciences*, edited by F. Clifford Rose and W. F. Bynum. New York: Raven Press, 1982.

Pinto-Correia, Clara. *The Ovary of Eve: Egg and Sperm and Preformation.* Chicago: University of Chicago Press, 1997.

Preuss, Julius. *Julius Preuss' Biblical and Talmudic Medicine.* Translated and edited by Fred Rosner. New York: Sanhedrin Press, 1978.

Reichman, Edward, and Fred Rosner. "The Bone Called Luz." *Journal of the History of Medicine and Allied Sciences* 51: 52–65.

Ruestow, E. G. "Leeuwenhoek's Perception of the Spermatozoa." *Journal of the History of Biology* 16: 185–24.

Schierbeek, A. *Measuring the Invisible World: The Life and Works of Antoni van Leeuwenhoek.* London and New York: Abelard-Schuman, 1959.

Terai, Takekazu. "Detection of Flatus Using a Portable Hydrogen Gas Analyzer." *Journal of Clinical Anesthesia* 15: Letter to the Editor (November 2003).

Zimmer, Carl. *Soul Made Flesh.* New York: Free Press, 2004.

Chapter 3: How to Weigh a Soul

Carpenter, Donald Gilbert. *Physically Weighing the Soul.* Online: www.1stbooks.com, 1998.

Clarke, John Henry. *A Dictionary of Practical Materia Medica*. London: Homeopathic Publishing Company, 1900-02.

Coulton, G. G. *From St. Francis to Dante: Translations From the Chronicle of the Franciscan Salimbene (1221–1288),* 2d ed. New York: Russell and Russell, 1968.

Haverhill Evening Gazette. "Weight of a Soul." 11 March 1907.

Hollander, Lewis E., Jr. "Unexplained Weight Gain Transients at the Moment of Death." *Journal of Scientific Exploration* 15 (4): 495–500.

Journal of the American Society for Psychical Research. Vol. 1, No. V (May 1907): Correspondence pages, pp. 263–83.

Kleiber, Max. *The Fire of Life: An Introduction to Animal Energetics.* Huntington, NY: Robert E. Krieger, 1975.

Macdougall, Duncan. "Hypothesis Concerning Soul Substance Together with Experimental Evidence of the Existence of Such Substance." *American Medicine* New Series Vol. II (4): 240–43 (April 1907).

New York Times. "Soul Has Weight, Physician Thinks." 11 March 1907, p. 5.

Sanctorius, Santorio. *De statica medicina: being the aphorisms of Sanctorius, translated into English*. Third edition, edited by John Quincy. London: W. and J. Newton, 1723.

Sunday Post (Boston). "Existence of 'Soul' Tested by Doctors." 10 March 1907.

Twining, H. LaV. *The Physical Theory of the Soul.* Westgate, CA: Published by the author, 1915.

Chapter 4: The Vienna Sausage Affair

Carrington, Hereward. *Laboratory Investigations into Psychic Phenomena.* New York: Arno Press, 1975.

——. *The Story of Psychic Science*. London: Rider, 1930.

Eisenberg, Henry. *Radiology: An Illustrated History.* St. Louis, MO: Mosby-Year Book, 1992.

Krauss, Rolf. *Beyond Light and Shadow: The Role of Photography in Certain Paranormal Phenomena.* Translated by Timothy Bill and John Gledhill. Munich: Nazraeli Press, 1992.

New York Times. "As To Picturing the Soul." 24 July 1911, p. 1.

Russ, Charles. "An Instrument Which Is Set in Motion by Vision." *Lancet*, 30 July 1921, pp. 222–24.

Sunday Post (Boston). "Heaven Is Perhaps Just Outside Earth." 21 May 1914.

Chapter 5: Hard to Swallow

Bird, J. Malcolm. "Our Next Psychic: A Preliminary Account of the Case that Now Comes Before Us, as It Appears to the Naked Eye." *Scientific American*, July 1924, p. 28.

Bondeson, Jan. *A Cabinet of Medical Curiosities.* New York: W. W. Norton, 1999.

Brockbank, E. M. "Merycism or Rumination in Man." *British Medical Journal*, 23 February 1907, pp. 421–27.

Crawford, W. J. *Experiments in Psychical Science.* New York: E. P. Dutton, 1919.

——. *The Psychic Structures at the Goligher Circle*. New York: E. P. Dutton, 1921.

Einhorn, Max. "Rumination in Man." *Medical Record*, 17 May 1890, pp. 554–58.

Fournier d'Albe, E. E. *The Goligher Circle (May to August, 1921), With an Appendix Containing Extracts from the Correspondence of the Late W. J. Crawford, D. Sc. And Others.* London: John M. Watkins, 1922.

Free, E. E. "Our Psychic Investigation: Preliminary Committee Opinions on the 'Margery' Case." *Scientific American*, November 1924, p. 304.

Gaskill, Malcolm. *Hellish Nell.* London: Fourth Estate, 2001.

Houdini, Harry. *A Magician Among the Spirits.* New York: Arno Press, 1972.

Jastrow, Joseph. "Ectoplasm, Myth or Key to the Unknown?" *New York Times*, 30 July 1922, p. 1.

New York Times. "'Ectoplasm' Prints Called Lung Tissue." 28 February 1926, p. 22.

New York Times. "Links Alchemists with Spiritualism." 14 April 1922, p. 14.

New York Times. "Man Bites a Ghost and Upsets Seance." 10 November 1923, p. 15.

New York Times. "Sorbonne Scientists Find No Ectoplasm After Experiments in Fifteen Seances." 8 July 1922, p. 1.

Popular Science Monthly. "Weighing Ghosts and Photographing Phantoms: How Three European Scientists Brought the 'Spirit World' into the Laboratory." September 1921, pp. 15–16.

Price, Harry. *Regurgitation and the Duncan Mediumship.* London: National Laboratory of Psychical Research, 1931.

Schrenck-Notzing, Albert von. *Phenomena of Materialization.* Reprint of 1920 ed., in the series Perspectives in Psychical Research. New York: Arno Press, 1975.

Wilson, William. "Rumination in Man." Letter to the editor in *Lancet*, 1839–40, pp. 671–72.

Chapter 6: The Large Claims of the Medium

Anonymous committee report. "Report on the Oliver Lodge Posthumous Test." *Journal of the Society for Psychical Research* 38 (685), pp. 121–43 (September 1955).

Findlay, James Arthur. *Looking Back: The Autobiography of a Spiritualist.* London: Psychic Press, 1955.

Hyman, Ray. "How *Not* to Test Mediums: Critiquing the Afterlife Experiments." *Skeptical Inquirer*, January/February 2003, pp. 20–30.

Matla, J. L. W. P., and G. J. Zaalberg van Zelst. *Le Mystère de la Mort*, 2d ed. Paris: G. Doin.

New York Times. "Detroit Student of Spirit Communication Ends Life, Perhaps in Effort to Test Theory." 7 February 1921, p. 1.

New York Times. "Owen Says Heaven Needs Active Men." 5 February 1923, p. 9.

Schouten, Sybo A. "An Overview of Quantitatively Evaluated Studies with Mediums and Psychics." *Journal of the American Society for Psychical Research* 88: 221–54 (July 1994).

Schwartz, Gary E. *The Afterlife Experiments: Breakthrough Scientific Evidence of Life After Death.* New York: Atria Books, 2003.

——. "Evidence of Anomalous Information Retrieval Between Two Mediums: Replication in a Double-Blind Design." *Journal of the Society for Psychical Research* 67 (2): 115–30 (April 2003).

Stevenson, Ian, Arthur T. Oram, and Betty Markwick. "Two Tests of Survival After Death: Report on Negative Results." *Journal of the Society for Psychical Research* 55 (815): 329–36 (April 1989).

Tyrrell, G. N. M. "The O. J. L. Posthumous Packet." *Journal for the Society for Psychical Research*, September 1948, pp. 269–71.

Chapter 8: Can You Hear Me Now?

Baruss, Imants. "Failure to Replicate Electronic Voice Phenomenon." *Journal of Scientific Exploration* 15 (3): 355–67.

Cooke, Andrew. "Electroplasm: Technology's Indissoluble Link to the Spirit World." Master's thesis, Royal College of Art, 2001.

Dusen, Wilson van. "The Presence of Spirits in Madness." Fourth ed. of pamphlet. New York: Swedenborg Foundation, 1983.

Ellis, D. J. *The Mediumship of the Tape Recorder: A Detailed Examination of the Phenomenon of Voice Extras on Tape Recordings*. Cambridge University Perrott-Warrick Fellowship (1970–72) report, published June 1978 (small-offset litho).

Fuller, John G. *The Ghost of 29 Megacycles*. London: Souvenir Press, 1985.

Johnson, Kristin, ed. *"Unfortunate Emigrants": Narratives of the Donner Party*. Logan, UT: Utah State University Press, 1996.

Lescarboura, Austin. "Edison's Views on Life and Death." *Scientific American*, 30 October 1920, p. 446.

Mason, D. H. "Psychic Psounds and Medium Tones: Spiritualism on 78." Series of three articles. *The Historic Record and AV Collector* 35: 30–34, 36: 17–20, 37: 16–24 (April, July, October 1995).

Ronell, Avital. *The Telephone Book: Technology, Schizophrenia, Electric Speech*. Lincoln: University of Nebraska Press, 1989.

Rousseau, David, and Julie Rousseau. "The Spellchecker Case." *Journal of the Society for Psychical Research*. (October 2005).

Runes, Dagobert D., ed. *The Diary and Sundry Observations of Thomas Alva Edison*. Westport, CT: Greenwood Press, 1968.

Tesla, Nikola. *My Inventions*. Williston, VT: Hart Brothers, 1982.

Watson, Thomas A. *Exploring Life: The Autobiography of Thomas A. Watson*. New York and London: D. Appleton, 1926.

Weightman, Gavin. *Signor Marconi's Magic Box*. Cambridge, MA: Da Capo Press, 2003.

Chapter 9: Inside the Haunt Box

Bruchard, J. F., D. H. Nguyen, and E. Block. "Effects of Electric and Magnetic Fields on Nocturnal Melatonin Concentrations in Dairy Cows." *Journal of Dairy Science* 81: 722–27 (1998).

MacDonald, Douglas, and Daniel Holland. "Spirituality and Complex Partial Epileptic-like Signs." *Psychological Reports* 91: 785–92 (2002).

Persinger, Michael A. "Average Diurnal Changes in Melatonin Levels Are Associated with Hourly Incidence of Bereavement Apparitions: Support for the Hypothesis of Temporal (Limbic) Lobe Microseizing." *Perceptual and Motor Skills* 76: 444-46 (1993).

——. "Increased Geomagnetic Activity and the Occurrence of Bereavement Hallucinations: Evidence for Melatonin-Mediated Microseizing in the Temporal Lobe?" *Neuroscience Letters* 88: 271–74 (1988).

——. "Experimental Facilitation of the Sensed Presence: Possible Intercalation between the Hemispheres Induced by Complex Magnetic Fields." *Journal of Nervous and Mental Disease* 190 (8): 533–41 (2002).

——, S. A. Koren, and R. P. O'Connor. "Geophysical Variables and Behavior: CIV. Power-Frequency Magnetic Field Transients (5 Microtesla) and Reports of Haunt Experiences Within an Electronically Dense House." *Perceptual and Motor Skills* 92: 673–74 (2001).

——, S. G. Tiller, and S. A. Koren. "Experimental Simulation of a Haunt Experience and Elicitation of Paroxysmal Electroencephalographic Activity by Transcerebral Complex Magnetic Fields: Induction of a Synthetic 'Ghost'?" *Perceptual and Motor Skills* 90: 659–74 (2000).

Randall, Walter, and Steffani Randall. "The Solar Wind and Hallucinations—A Possible Relation Due to Magnetic Disturbances." *Bioelectromagnetics* 12: 67–70 (1991).

Chapter 10: Listening to Casper

Altmann, Jürgen. "Acoustic Weapons: A Prospective Assessment." *Science and Global Security* 9: 165–234.

Davis, Laura. "Soundless Concert Stirs the Emotions." *Daily Post* (Liverpool), 17 February 2003.

Muggenthaler, Elizabeth von. "Low Frequency and Infrasonic Vocalizations from Tigers." Paper 3aABb1, presented at the annual meeting of the Acoustical Society of America/NOISE-CON, Newport Beach, CA, 2000.

Tandy, Vic, and T. R. Lawrence. "The Ghost in the Machine." *Journal of the Society for Psychical Research* 62: 360–64 (1998).

——. "Something in the Cellar." *Journal of the Society for Psychical Research* 64 (3): 129–40 (July 2000).

Walsh, Edward J., et al. "Acoustic Communications in *Panthera tigris*: A Study of Tiger Vocalization and Auditory Receptivity." Paper 4aAB3 presented at the annual meeting of the Acoustical Society of America, Nashville, TN, May 2003.

Chapter 11: Chaffin v. the Dead Guy in the Overcoat

"Case of the Will of James L. Chaffin." Editor's report in *Proceedings of the Society for Psychical Research* 36: 517–24 (1928).

Cornell, A. D. "An Experiment in Apparitional Observation and Findings." *Journal of the Society for Psychical Research* 40 (701): 120–24.

——. "Further Experiments in Apparitional Observation." *Journal of the Society for Psychical Research* 40 (706): 409–18.

Davie County Enterprise Record. "Dead Man Returns in a Dream; An Estate in Davie is Redivided." 4 January 1979, p. 5B.

Gurney, Edmund, F. W. H. Myers, and Frank Podmore. *Phantasms of the Living*. New York: E. P. Dutton, 1918.

Osborn, Albert S. *Questioned Documents*, 2d ed. Albany, NY: Boyd Printing, 1929.

Wall, James W. *History of Davie County*. Spartanburg, SC: Reprint Publishing, 1997.

Chapter 12: Six Feet Over

Atwater, P. M. H. "Is There a Hell? Surprising Observations About the Near-Death Experience." *Journal of Near-Death Studies* 10 (5): 149–60.

Becker, Carl. "The Pure Land Revisited: Sino-Japanese Meditations and Near-Death Experiences of the Next World." *Anabiosis—The Journal for Near-Death* 4 (1): 51–68 (Spring 1984).

Blackmore, Susan. "Near-Death Experiences: In or Out of the Body?" *Skeptical Inquirer* 16: 34–45 (Fall 1991).

Blanke, Olaf, et al. "Stimulating Illusory Own-Body Perceptions." *Nature* 419: 269 (September 2002).

Cheek, David. "The Anesthetized Patient *Can Hear* and *Can Remember*." *American Journal of Proctology* 13 (5): 287–89 (October 1962).

Clark, Kimberly. "Clinical Interventions with Near-Death Experiences." In *The Near-Death Experience: Problems, Prospects, Perspectives.* Springfield, IL: Charles C. Thomas, 1984.

Cook, Emily Williams, Bruce Greyson, and Ian Stevenson. "Do Any Near-Death Experiences Provide Evidence for the Survival of Human Personality after Death? Relevant Features and Illustrative Case Reports." *Journal of Scientific Exploration* 12 (3): 377–406 (1998).

Greyson, Bruce, and Nancy Evans Bush. "Distressing Near-Death Experiences." *Psychiatry* 55: 95–110.

Jansen, Karl. *Ketamine: Dreams and Realities.* Sarasota, FL: Multidisciplinary Association for Psychedelic Studies, 2001.

Morris, Robert L., et al. "Studies of Communication During Out-of-Body Experiences." *Journal of the American Society for Psychical Research* 72 (1): 1–21 (January 1978).

Osis, Karlis, and Donna McCormick. "Kinetic Effects at the Ostensible Location of an Out-of-Body Projection During Perceptual Testing." *Journal of the American Society for Psychical Research* 74 (3): 319–29 (July 1980).

Parnia, S., et al. "A Qualitative and Quantitative Study of the Incidence, Features, and Aetiology of Near-Death Experiences in Cardiac Arrest Survivors." *Resuscitation* 48 (2): 149–56 (February 2001).

Ring, Kenneth. "Further Evidence for Veridical Perception During Near-Death Experiences." *Journal of Near-Death Studies* 11 (4): 223–29 (Summer 1993).

——, and Sharon Cooper. *Mindsight: Near-Death and Out-of-Body Experiences in the Blind.* Palo Alto, CA: William James Center for Consciousness Studies, 1999.

Sabom, Michael. *Recollections of Death: A Medical Investigation.* New York: Harper & Row, 1982.

——. "The Shadow of Death." Parts 1 and 2. *Christian Research Journal* 26 (2): 14–21 and 26 (3): 45–51 (2003).

Schwender, D., et al. "Conscious Awareness During General Anesthesia: Patients' Perceptions, Emotions, Cognition, and Reactions." *British Journal of Anesthesia* 80: 133–39 (1998).

Van Lommel, W. "About the Continuity of Our Consciousness." In *Brain Death and Disorders of Consciousness.* Edited by C. Machado and D. A. Shewmon. New York, Boston, Dordrecht, London, Moscow: Kluwer Academic/ Plenum Publishers, 2004.

——, et al. "Near-Death Experience in Survivors of Cardiac Arrest: A Prospective Study in the Netherlands." *Lancet* 358: 2039–45 (2001).

Mary Roach is the author of *Stiff*. Her writing has appeared in *Salon*, *Wired*, *Outside*, *GQ*, *Discover*, *Vogue*, and the *New York Times Magazine*. She lives in Oakland, California.